Patrick Moore's Practical Astronomy Series

P9-AQN-473

JUL 02 2004

Springer
London
Berlin
Heidelberg
New York
Hong Kong
Milan
Paris
Tokyo

Other Titles in this Series

Observing and Measuring Visual Double Stars

Bob Argyle (Ed.)

With 106 Figures

Springer

QB
821
.O2
2004
4CD-Rom

British Library Cataloguing in Publication Data
Observing and measuring visual double stars. – (Patrick
    Moore's practical astronomy series)
    1. Double stars – Observers' manuals 2. Double stars –
    Observations
    I. Argyle, Bob
    523.8'41
    ISBN 1852335580

Library of Congress Cataloging-in-Publication Data
Observing and measuring visual double stars / Bob Argyle (ed.).
    p. cm. – (Patrick Moore's practical astronomy series,
    ISSN 1617-7185)
    Includes bibliographical references.
    ISBN 1-85233-558-0 (acid-free paper)
    1. Double stars–Observers' manuals. I. Argyle, Robert W. II. Series.
QB821.O2 2004
523.8'41–dc22   2003061881

Apart from any fair dealing for the purposes of research or private
study, or criticism or review, as permitted under the Copyright,
Designs and Patents Act 1988, this publication may only be reproduced, stored or transmitted, in any form or by any means, with the
prior permission in writing of the publishers, or in the case of repro-
graphic reproduction in accordance with the terms of licences issued
by the Copyright Licensing Agency. Enquiries concerning reproduction outside those terms should be sent to the publishers.

Patrick Moore's Practical Astronomy Series ISSN 1617-7185
ISBN 1-85233-558-0 Springer-Verlag London Berlin Heidelberg
Springer-Verlag is a part of Springer Science+Business Media
springeronline.com

© Springer-Verlag London Limited 2004
Printed in the United States of America

The software disk accompanying this book and all material
contained on it is supplied without any warranty of any kind.

The use of registered names, trademarks, etc. in this publication does
not imply, even in the absence of a specific statement, that such
names are exempt from the relevant laws and regulations and therefore free for general use.

The publisher makes no representation, express or implied, with
regard to the accuracy of the information contained in this book and
cannot accept any legal responsibility or liability for any errors or
omissions that may be made. Observing the Sun, along with a few
other aspects of astronomy, can be dangerous. Neither the publisher
nor the author accepts any legal responsibility or liability for personal loss or injury caused, or alleged to have been caused, by any
information or recommendation contained in this book.

Typeset by EXPO Holdings, Malaysia
58/3830-543210 Printed in acid-free paper SPIN 10844341

528600220

Acknowledgements

I am extremely grateful to my colleagues who have contributed their expertise so willingly in the chapters within: Andreas Alzner, Graham Appleby, Owen Brazell, Michael Greaney, Andreas Maurer, Michael Ropolewski, Christopher Taylor, Tom Teague, Nils Turner and Doug West. I also thank Andreas Alzner, Michael Greaney and Jean-François Courtot for help in proof-reading. Needless to say, mine is the final responsibility for any errors which might escape the various proof-reading exercises.

My thanks are also due to Patrick Moore who first suggested the idea of this volume and John Watson, Springer's Managing Editor, who has been very supportive throughout the whole process.

Finally my wife Angela has not only had to contend with many hours of my sitting in front of the computer but has actively encouraged me to "get the thing finished".

Bob Argyle
Waterbeach, Cambridgeshire
2003 March

Zum Raum wird hier die Zeit (Time here becomes space) – Parsifal, Act 1.

Contents

Introduction

Double stars are the rule, rather than the exception, in the solar neighbourhood and probably beyond. Current theories of star formation point to multiple stars or stars and planets as the preferential outcome of gravitating protostellar material. Stellar pairs can be detected at many wavelengths from X-rays, where modern satellites can resolve the two brightest components of Castor (separation 3.8″), to the radio where the precision of long baseline interferometry can also see the 4 milliarcsecond (mas) "wobble" in the 2.87-day eclipsing system of Algol and can distinguish which of the two stars is emitting the radio waves. They come in a wide range of orbital sizes, periods and masses. From Groombridge 34, where the stars are separated by five times the distance of Pluto from the Sun and whose motion is barely perceptible, through the spectroscopic binaries with periods of weeks, down to exotic pairs like double white dwarf contact systems with periods of 5 minutes. From young O-star binaries like R136–38 in the Large Magellanic Cloud containing two extremely bright and hot stars, of 57 and 23 solar masses, down to the snappily named 2MASS J1426316+155701, a pair of brown dwarfs with masses only 0.074 and 0.066 times that of the Sun.

In this volume we are concentrating on only one aspect, the visual double stars, which can define as those pairs which can be seen or imaged in a telescope of moderate aperture. The classic image of the double-star observer as a professional scientist with a large refractor and a brass filar micrometer is no longer valid. Researchers cannot afford to spend a lifetime measuring a large number of pairs in order to get a few dozen orbits. The high precision astrometric satellites, ground-based interferometer arrays, and infrared speckle interferometry have all helped, respectively, to discover large numbers of new pairs, push direct detection into the spectroscopic regime with measurement of binaries with periods of a few days, and to probe the near and mid-infrared where faint red and brown dwarf companions and, ultimately, planets appear. This has left a large number of wide, faint pairs which are under-observed.

There has been a common perception that double-star observing is either not very interesting or does not afford any opportunities for useful work. The aim of this book is to dispel these views and indicate where observers might usefully direct their efforts. At the basic level, we give advice about how to observe them with binoculars and small telescopes. At a more serious level, chapters about micrometers, CCD cameras and other techniques have been included. For those who do not wish to spend several hundred pounds on a filar micrometer, the graticule eyepieces such as the Celestron Micro Guide available for catadioptric tele-scopes can be used effectively for relative position mea-surement of wider pairs, and for those who find observing too taxing, astrometry of faint pairs can be done by examination of some of the huge catalogues produced from the various Schmidt surveys.

Clearly, for the observer, the role of the telescope is very important. For casual viewing any optical aid can give reasonable views of wide and bright double stars. I spent several years accumulating visual estimates of colour, magnitudes, and relative positions of more than 1000 pairs using a 21-cm reflector using Webb's *Celestial Objects for Common Telescopes (Volume II)* and *Norton's Star Atlas* (15th edition, 1964). Even in *Norton's* many of the measures given were more than 30 years old and it was this that sparked an interest in obtaining a micrometer to bring them up-to-date.

On the whole, equatorially mounted telescopes are almost a necessity and although Dobsonian telescopes can give fine views of double stars, using them for mea-surement is not straightforward. Potential users should look at Chapter 22 where Michael Greaney shows how to calculate position angle in situations where the field rotates. Whilst the grating micrometer (described by Andreas Maurer in Chapter 14) is relatively insensitive to the lack of an RA drive the field rotation is an added problem.

Resolution is ultimately dependent on aperture and although many of the most interesting binaries are significantly closer than 1″ the aperture available to today's observers is no longer limited to the small sizes that were common about 30 or 40 years ago when the 12.5-inch reflector was the exception rather than the rule. These days no one is surprised to see amateur observers sporting 20, 30 or even 40-inch telescopes and for those who thought that refractors were the required telescope for double-star observing Christopher Taylor has other ideas.

In the last 10 years the CCD camera has become a dominant force in observational astronomy. As both a positional and photometric detector it has excellent applications in the observation of double stars and these will be discussed later by Doug West.

Filar micrometers are available commercially, costing about half the price of a CCD camera and usable up to the resolving limit of the telescope. The human eye is still the best all-round detector available for work on close pairs of images whether they are equally or unequally bright.

Those with the larger apertures, however, should consider the speckle interferometer as an alternative to the micrometer. With atmospheric effects becoming more significant with telescope size, the speckle camera can punch through the turbulence and produce diffraction-limit imaging. Nils Turner describes how this can be achieved at relatively low cost.

The availability of inexpensive and yet powerful personal computers has brought several other aspects of double-star astronomy within reach. The latest static version of the United States Naval Observatory (USNO) double star catalogue, WDS 2001.0, is now available on CD-ROM (the regularly updated WDS catalogue is available on-line only and incremental files can be downloaded to update the static version of the catalogue). It is no longer necessary to measure the bright pairs which appear in the popular observing guides. With the WDS the more neglected pairs can be selected for measurement and charting software makes finding even the most obscure pair much easier. The USNO have placed on their website several lists of neglected double stars which they would like observers either to confirm as double or to make new measures. Many of the catalogues available on the WDS CD-ROM can also be found on the CD-ROM available with this volume.

Orbital computation, once the province of specialists, can now be done by anyone but it is not to be taken lightly. Even if all the measures of a particular system can be rounded up it still requires an appreciation of the quality of the observations and the existence of systematic errors. How do you combine measures by Struve in 1828 with those by van

Biesbroeck in 1935 and speckle measures made in 1990? Perhaps most importantly is a new orbit necessary and is yours better than any others? Andreas Alzner has contributed two chapters on this important topic.

Finally, what about the double stars themselves? As we have seen, current research is pushing resolution to unprecedented limits but in the meantime who is paying any attention to the 90,000 plus pairs in the Washington Double Star (WDS) catalogue, the central repository for the subject? In particular, who is watching the southern binaries, many of which are being overlooked? I recently found four systems in the WDS catalogue which did not have orbits, one of which, δ Velorum, is 2nd magnitude. Its 5th magnitude companion was not observed for 50 years and has recently passed through periastron. Thanks to Andreas Alzner, orbits for these pairs have now been computed, but confirming observations are also needed.

Chapter 1

More Than One Sun

Bob Argyle

Introduction

On a clear, dark night several thousand stars can be seen at any one time. They form familiar patterns such as the Great Bear and Cygnus in the northern hemisphere and Scorpio and Crux in the south. The distances are so great that we see the constellation patterns essentially unchanged from those seen by the ancient Egyptians, for instance. This is partly due to the fact that some of the bright stars in constellations are in what are called moving groups – a loose association of stars moving through space together. More tightly bound are clusters of star such as the Pleiades or Seven Sisters which appears in the northern sky in the late summer. Eventually the moving groups and clusters of stars will gradually disperse because the distance between the stars is such that the gravitational attraction between the members is relatively weak.

Those with keen eyes will be able to see some close pairs of stars without optical aid. The most famous is Mizar and Alcor in the tail of the Great Bear. The first recorded "naked-eye" pair is ν Sgr which was mentioned by Claudius Ptolemy in his famous *Almagest* catalogue of *c.* AD 140. It is described[1] as "The star in the middle of the eye (of Sagittarius) which is nebulous and double". The angular separation of this pair is 13′, or about the same separation as Mizar and Alcor. As a comparison, the apparent diameter of the Full Moon is 30′.

Relative Positions in Visual Double Stars

The separation is one of two quantities needed to fully describe the relative position of double stars, the other being the position angle. With the brighter of the two stars being taken as the origin, the separation is defined as the angular distance in arcseconds between the two stars and the position angle is the bearing of the fainter star from the brighter in degrees with north being taken as 0 degrees, east is 90 degrees, and so on. Figure 1.1a shows the situation for the naked eye and binoculars. When a telescope is used the view is inverted so Figure 1.1b applies to telescopic views.

It is usual to represent separation by the Greek letter rho (ρ) and position angle by the Greek letter theta (θ). These terms will be used throughout this book. Another common term is Δm which is shorthand for the difference of magnitude between the primary and secondary stars. Unless otherwise stated the magnitudes in this book will be visual. The fainter of the two stars is sometimes called the *comes*, a Latin word meaning companion.

Figure 1.1.
a Telescopic view;
b naked eye and binoculars.

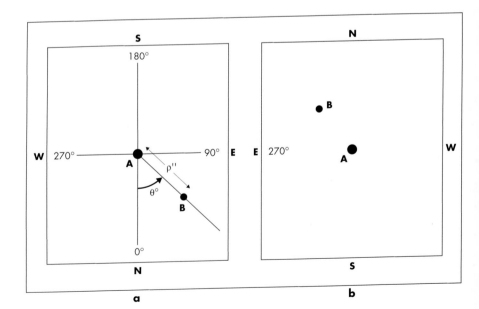

a

b

Naked Eye Limits

In the case of the human eye, the closest pair of stars which can be seen unaided depends on the diameter of the pupil. This, in turn, depends on the lighting conditions and when fully dark-adapted the pupil may be 6 or 7 mm in diameter, suggesting that the limit of resolution from the Airy formula (see Chapter 10) is about 20″ but the presence of aberrations in the eye and the low light levels from the night sky conspire to reduce the effective resolution to about 2.5′.

In practice a normal pair of eyes should be able to see the stars θ^1 and θ^2 Tauri at 5.5′ without difficulty and some may be able to make out ε^1 and ε^2 Lyrae at 3.6′. The ability to resolve naked-eye pairs tends to deteriorate with age, and younger eyes will probably do better, although practice undoubtedly enhances keenness of vision. Sight, like hearing, or any of the five senses, can be improved with experience. Table 1.1 contains a short list of bright wide pairs which, it is suggested, can be used as a test of naked-eye resolving power. In some of the cases, both stars have a Bayer letter or Flamsteed number and these are used as the main identifier. The positions are given for equinox 2000.0 together with the date of the most recent measure, the visual magnitudes of both stars and the position angle and separation of the pair. Most of these pairs are the results of chance alignment.

Table 1.1. Some naked-eye double stars.

RA 2000	Dec 2000	Pair	Epoch	PA (°)	Separation (″)	V_a	V_b
0318.2	−6230	ζ Ret	1991	217	309.3	5.24	5.33
0425.4	+2218	κ Tau	1991	173	340.2	4.21	5.27
0428.7	+1552	θ Tau	1991	347	337.2	3.40	3.84
0439.3	+1555	σ Tau	1991	194	437.5	4.67	5.08
0718.3	−3644	Jc 10 Pup	1991	98	240.0	4.65	5.11
1208.4	−5043	δ Cen	1991	325	269.1	2.58	4.46
1450.9	−1603	α Lib	1991	314	230.7	2.75	5.15
1622.4	+3348	ν CrB	1984	165	359.5	5.20	5.39
1844.3	+3940	ε Lyr	1998	174	210.5	4.59	4.67
1928.7	+2440	6–8 Vul	1991	28	422.9	4.44	5.82
2013.6	+4644	31–32 Cyg	1998	324	330.7	3.80	4.80
2018.1	−1233	α Cap	1984	290	381.4	3.80	4.20

Optical Pairs

Optical double stars are simply formed due to line-of-sight coincidence. They are usually widely separated (> 5″ or so) and the proper motions, or the individual motions in right ascension and declination, of each component, across the sky, are significantly different. In addition, the stars are usually unequally bright, reflecting the difference in distances but this by itself is not a criterion. A good example is δ Herculis where the two stars were separated by more than 34″ at discovery by the elder Herschel in 1779, they closed up to about 8″.8 in 1964 and are now at 11″ and widening (Figure 1.2a). Such pairs are usually of no direct scientific interest to astronomers but can produce some fine sights in small telescopes. The stars in δ Herculis are, for instance, pale yellow and blue in colour and the primary is about 24 parsecs distant. Little is known about the companion.

Telescopic Pairs

Whilst binoculars, particularly the image-stabilised variety (see Chapter 3), can show literally hundreds of double stars, the use of a small telescope will considerably increase the number of pairs of stars that can be seen. It also allows the user to see stellar colours more

Figure 1.2. a The proper motion of δ Herculis. Measurements of the position angle and separation of star B with respect to A over many years shows the relative motion between the two. **b** shows the real situation with star A moving towards PA 187° at a rate of 0.159″ per year whilst B moves towards PA 275° by 0.117″ per year.

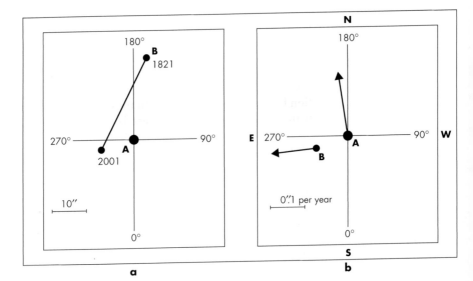

a   b

easily. In a 90-mm telescope, most of the closest pairs that can be seen are binary pairs – the two stars are physically connected by a mutual gravitational bond – and they rotate around the common centre of gravity in periods ranging from a few tens to a few millions of years.

Binary Stars

Visual Binaries

In the case of physically connected pairs of stars, what observers see when they plot the position angle and separation of the pair over a number of years is a curve. If followed for the whole orbital period the result would be an ellipse – this is the apparent orbit, in other words, the projection of the true orbit onto the plane of the sky. With a small telescope, hundreds of binary stars can be observed and of these the more nearby pairs offer the best chance of seeing the orbital motion over a few years. Estimates of separation can be made in terms of the diameter of the apparent disk of the brighter component which can be calculated for any telescope aperture using the Airy formula in Chapter 10. Position angle can be estimated to perhaps the nearest 5 or 10 degrees by eye by allowing the pair in question to drift through the field at high magnification with the driving motor stopped.

True (and apparent) orbits come in all shapes and sizes from circular to elongated ellipse but the tilt of the orbital plane can also vary from 90° (in which the plane is in the line of sight) to 0° in which we see the orbit face-on. To describe the real orbit fully requires seven quantities of which eccentricity, e, and inclination have just been explained. In the ellipse, the time at which the two stars are closest is called periastron (similar to perihelion when the Earth is nearest the Sun). The other values are the orbital period, P, in years (the time taken between successive arrivals by star B at the periastron point) and three values which describe the size and orientation of the orbit which are described fully in Chapter 7. The motion of star B around A follows Kepler's laws and in an exact analogy with the Solar System, the mass of both stars is related to the size of the orbit and the orbital period.

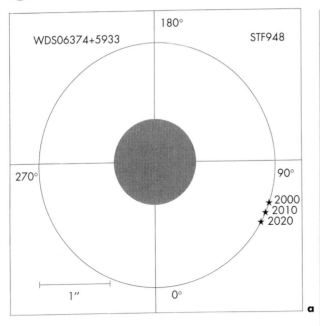

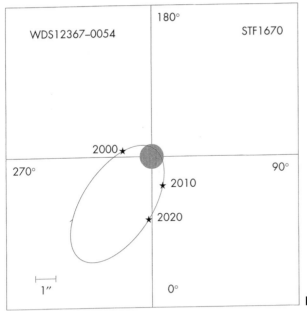

Figure 1.3. a The visual binary 12 Lyncis. $P = 706$ years, $e = 0.03$ and orbit inclined at 2° to the plane of the sky. **b** γ Virginis, $P = 169$ years, $e = 0.89$, inclined at 32° to the plane of the sky. The radius of the central circle indicates the Dawes limit for a 20-cm aperture. 12 Lyncis is therefore always visible in this aperture but γ Virginis will close to less than 0″.4 in early 2005 and will need at least 30 cm. At this time the position angle will change by 1° every five days!

Figure 1.3 gives an example of two well-known visual binaries. Contrast the orbital motion in both pairs by comparing the positions at 2000, 2010 and 2020.

To measure the total mass of both stars requires the apparent orbit to be defined as accurately as possible.

This can be done by measuring ρ and θ at different times, for as much of the orbit as is practical. (Long periods will mean that only a preliminary orbit can be obtained.) There are measuring techniques of various kinds which can be employed to accurately measure the relative position of B and to determine the values of ρ and θ. Later in this book the various methods that are available to the observer are mentioned in more detail.

For visual binaries, observations of the apparent orbit lead to the determination of the true orbit from which we can derive the sum of the masses, in terms of the solar mass, provided that the parallax is known. The astrometric satellite *Hipparcos* has been instrumental in providing parallaxes of high accuracy for a large number of binary stars.

Once we know the apparent orbit of a visual binary, we can, if the parallax of the system is also known, obtain the sum of the masses of the stars in the system via Kepler's third law:

$$M_1 + M_2 = \frac{a^3}{\pi^3 P^2} \qquad (1.1)$$

where a is the semimajor axis of the apparent ellipse, and π is the parallax. Both are in arcseconds and P is in years. The mass sum is then given in units of the Sun's mass.

To obtain the individual masses requires defining the apparent orbit for each component by measuring its position with time against the background field stars. The apparent orbits are identical with the relative sizes determining the ratio of the masses, the primary star, being the most massive, traces out the smaller ellipse (see Figure 7.1 in Chapter 7). Unfortunately this method only applies to a small number of wide, nearby pairs which can be resolved photographically throughout the orbit. The apparent orbits are identical in shape, with the relative size of each orbit being inversely proportional to the mass of the star:

$$\frac{M_1}{M_2} = \frac{a_2}{a_1} \qquad (1.2)$$

Combining (1.1) and (1.2) allows us to get the mass of each component.

The USNO Sixth Catalogue of Orbits[2] contains more than 1700 orbits of which 1433 refer to pairs resolvable by conventional techniques. Of these orbits, about 4% are grade 1, the longest period being that of 70 Oph at

Table 1.2. Distribution of orbit quality in the *USNO Sixth Catalogue of Orbits.*

Grade	Category	Longest period (years)	Number of pairs	Percentage of catalogue
1	Definitive	88.38	52	3.6
2	Good	257	198	13.8
3	Reliable	522.16	292	20.4
4	Preliminary	18212.2	454	31.7
5	Indeterminate	32000	437	30.4

88.38 years. Table 1.2 shows the distribution of the five main orbit grades. Throughout this volume reference will be made to the fifth and sixth editions of this catalogue. The fifth edition is available from the USNO on CD-ROM (see the appendix) whilst the sixth is the dynamic version which is regularly updated, but a copy of this version appears on the CD-ROM accompanying this book.

Spectroscopic Binaries

These are stars which appear single in all telescopes but turn a spectroscope on them and the spectral lines are observed to shift periodically with time due to the Doppler shift as the stars approach and then recede from the observer. The lines merge when the stars are both moving across the line of sight. There are two main types. When the stars are of similar brightness then two sets of spectral lines can be seen, particularly when one star is moving towards us and the other is moving away. These are called double-lined systems. When one star is much brighter than the other then only the spectral lines of the bright star can be seen to move periodically. This is called a single-lined system. Spectroscopic binaries have periods ranging from hours to a few tens of years. In a few rare cases they can also be resolved using speckle or ground-based interferometry. Such systems are important as they allow many characteristics of the component stars to be determined.

Astrometric Binaries

Again, these are single objects in all telescopes but reveal their duplicity by the effect that the unseen companion star has on the proper motion or the transverse motion of the star against the background of fainter stars. This motion will be constant for a single star but the presence of a companion constantly pulls on the primary star and the effect is to observe the star "wobble" across the sky. This was first noticed by Bessel in the proper motion of Sirius – some 3.7″ every year and large enough to be seen by regular measurement with respect to the neighbouring stars. Bessel rightly attributed the periodic wobble of Sirius to the presence of an invisible but massive companion. In 1862 Alvan Clark saw Sirius B for the first time, thus confirming Bessel's prediction.

Multiple Stars

Less common in the telescope, but more spectacular and worth seeking out, are the multiple stars. Systems like β Mon, with its three pure-white gems within 7″, ζ Cancri, of which more later, and ι Cas (yellowish, bluish and bluish, according to Robert Burnham).

If multiple stars are to be stable over a long time scale then they need to follow a certain hierarchy. In the case of a single star orbiting a close pair, the ratio of the orbital periods of the outer star around AB to that of the inner orbit AB is usually at least 10:1. This appears to apply from periods of about ten days up to thousands of years.

Quadruple stars, of which the most famous is the "double-double", epsilon Lyrae can be ordered in two ways. Firstly, as in epsilon's case, there are two pairs each orbiting the common centre of gravity. Alternatively, a double star is orbited by a distant third star and then even more distantly a fourth star circles the whole group.

Systems of higher multiplicity are known – perhaps the most famous is the sextuple system Castor, which is described in more detail in Chapter 9. A recent catalogue of multiple stars[3] lists 626 triples, 141 quadruples, 28 quintuples and ten sextuples. The existence of two systems thought to be septuple (ν Scorpii and AR Cas) awaits confirmation of further suspected components.

The Trapezium, which to a small telescope user is four stars embedded in the Orion Nebula, is the prototype of another sort of multiple star. It is not strictly ordered like the quadruples such as epsilon Lyrae, but is more a loose aggregation and can be regarded more as a small star cluster than a multiple star as such. It is none the less beautiful for this and seen against the glowing green background of the nebula, on a cold winter's night in a good telescope, it is one of the sky's most spectacular sights.

History of Double Star Observation

In 1610 the invention of the telescope by Galileo gradually led to the discovery of telescopic double stars but these were noted merely by the way. In 1617 Castelli found that Mizar was itself double[4] and he later added a few more pairs. In 1664 Robert Hooke was observing the comet discovered by Hevelius when he came across γ Arietis, a pair of pure-white stars of 4th magnitude separated by some 8″.

Over the next hundred years or so a few more double stars were noted but not catalogued in any determined manner, but this was to change when the Reverend John Michell first suggested that double stars were not merely a line-of-sight effect but that the two components really revolved around each other under a mutual gravitational influence, implying that Newton's laws applied to objects outside the Solar System. In *Philosophical Transactions* for 1767, Michell says: "it is highly probable in particular, and next to a certainty in general, that such double stars, &c, as appear to consist of two or more stars placed close together, do really consist of stars placed near together, and under the

Table 1.3. The first ten telescopic double star discoveries

Pair	Discovery	By
ζ UMa	1617 Jan	Castelli
β Mon	1617 Jan 30	Castelli
θ Orionis ABC	1617 Feb	Galileo
β Sco	1627	Castelli
γ Ari	1664	Hooke
Castor	1678?	G.D. Cassini
ζ Cnc AB-C	1680 Mar 22	Flamsteed
α Crucis	1685	Fontenay
α Centauri	1689	Richaud
γ Virginis	1718	Bradley

influence of some general law, whenever the probability is very great, that there would not have been any such stars near together, if all those that are not less bright than themselves had been scattered at random throughout the whole heavens".

A small catalogue of double stars was compiled in 1780 by Christian Mayer of Mannheim but the next great step was taken by William Herschel who turned his unprecedently powerful telescopes on many bright stars to find that even at high power, some stars appeared as very close pairs. In an attempt to measure stellar parallax, Herschel argued that in unequally bright, close pairs by measuring the position of the faint (hence distant and fixed) star with respect to the bright (or nearby) star he should be able to measure the parallactic shift and hence the distance of the latter. This idea he attributes to Galileo. To prove this he used filar micrometers of his own construction to measure the position of the fainter star with respect to the brighter. However, instead of seeing a six-monthly "wobble" in the position of the bright star with respect to the faint, Herschel found that the relative motion between the two stars was curved and could only be explained if the stars were revolving around a common centre of gravity. He had proved that binary stars existed but the mathematical confirmation came six years after his death, in 1828, when the French scientist Savary used the pair ξ UMa (which Herschel had discovered) to show that the apparent orbit of the fainter star around the brighter (assuming the latter was fixed) was an ellipse.

The significance of this work was that it gave an estimate for the ratio of the stellar masses in a binary star system. This resulted in a great impetus in the visual observation of double stars and over the next 50 years or so many rich amateur astronomers in Europe dedicated time and money to making micrometric measurements, or paid someone to do it for them. Dawes, in England, and particularly Baron Ercole Dembowski, in Italy, and others, flourished but without the excitement of discovery the work lost momentum and became largely unfashionable by the turn of the century.

In 1857 when Bond first imaged Mizar with the Harvard 15-inch refractor the advantages of photography for double-star astronomy were not immediately realised, partly because the resolution obtained initially did not allow much work to be done in the orbital pairs of relatively short period. For those bright pairs where the separation was such that both components could be imaged at all parts of the orbital cycle, such as 70 Oph, it was possible to determine individual masses from the size of the apparent ellipses that each star traced out against the stellar background. It was not until the middle of the last century that observers such as Willem Luyten, Peter van de Kamp and Wulff Heintz used photography much more purposefully. Luyten, in a long career, found many pairs of stars with common proper motion, indicative of orbital pairs but with a long period. van de Kamp concentrated on those systems where the only evidence of duplicity was a periodic wobble of a bright star with respect to the background, indicating a faint and close but nonetheless significantly massive companion star.

The Great Era of Discovery

From 1870 or so, when the American astronomer S.W. Burnham first started in double-star astronomy, a golden period for discovery opened up and continued for about 80 years, first in the northern hemisphere and latterly in the south. The largest refractors in existence were used in systematic surveys of the BD star catalogues by R.G. Aitken and W.J. Hussey in California (they discovered 4700 pairs between them) and some years later by R.T.A. Innes, W.H. van den Bos and W.S. Finsen at the Republic Observatory, Johannesburg (5000 discoveries) and Rossiter and colleagues at the Lamont–Hussey Observatory at Bloemfontein (7650 discoveries) in South Africa. When the latter retired in 1952 it was not long before P. Couteau and P. Muller in France began to search for new pairs again, dividing up the northern heavens with Couteau tackling the zones from +17° to +52° and Muller surveyed the zones near the north pole. They were remarkably successful and Couteau's list now exceeds 2700 new pairs whilst Muller found more than 700. Additionally, W.D. Heintz has detected 900 new pairs, most of them in a zone close to the equator and in the southern hemisphere.

Modern Techniques

Although it was proposed by Albert Michelson almost a hundred years ago, stellar interferometry is today even more important as a means of researching the dynamics of binary stars as it was then. Michelson's idea led to the construction of an interferometer for the 100-inch reflector on Mount Wilson in the 1920s, consisting of a 20-foot structure with flat mirrors at each end mounted at the top end of the telescope tube.

This instrument uses the interference of light to determine whether a bright single star is either extended i.e. its diameter is resolvable at the Earth, or a close double. By combining the light from each of the two small mirrors and adjusting the separation of the mirrors until the fringes thus formed combined in such a way that they cancelled each other out then the separation of the two components could be found from the separation of the mirrors and the position angle from the orientation of the fringes. With so little light available only bright stars could be measured.

In 1925 Frederick Pease[5] first resolved Mizar A using this equipment. It was also used for observations of extended sources, so that, for instance, the diameters of supergiant stars such as Betelgeuse could be determined. Other stars measured included the binary system Capella which turned out to have a separation of between 0.03″ and 0.05″ and a period of 104 days.

In the 1970s double-star observation underwent a revolution with the invention of speckle interferometry (see Chapter 17). This technique effectively removes the effect of the atmosphere and allows telescopes to operate to the diffraction limit. In the case of the 4-metre reflectors on which it was used, this corresponded to about 0″.025 or about four times closer that Burnham or Aitken could measure. In addition the accuracy of this method was much greater than visual measures and since then it has proved its worth by discovering new very close and rapid binaries and improving the older visual orbits.

The launch of the *Hipparcos* satellite in 1989 also heralded a new era of double-star discovery. Operating high above the atmosphere its slit detectors found some 15,000 new pairs, most of which are difficult objects for small telescopes but a number have already been picked up in very small apertures.

The Future

Where does double-star observation go next? In the immediate future it will be from the ground where a number of specially built optical arrays will be operating over the next few years.

At Cambridge in the UK, the COAST (Cambridge Optical Aperture Synthesis Telescope) five-mirror interferometer has been working for some years with a current baseline of 48 metres and there are plans to extend this to 100 metres. This is an extension of the Michelson instrument at Mount Wilson. By using more mirrors and using the Earth's spin to rotate the instrument with respect to the star, astronomers have used phase closure, a technique first used in radio astronomy, to effectively image the structure of stars such as Betelgeuse.

It has easily resolved the bright spectroscopic binary Capella, whose components are about 50 milliarcseconds (mas) apart. Another such instrument, the NPOI (Navy Prototype Optical Interferometer) using 50-metre baselines in Arizona, has resolved spectroscopic binaries such as the brighter component of Mizar. Long known to have a period of 20.5 days, the NPOI can detect and measure the individual stars even though at closest approach they are only 4 mas apart (see Figure 9.1). The combination of the NPOI data and the spectroscopic data can give very accurate values for the size of the orbit, the parallax of the system and the individual masses, and the radius of each component.

Soon the CHARA (Centre for High Resolution Astronomy, Georgia State University) array in Arizona which employs 1-metre telescopes will be operating with a 350-metre baseline, and the Sydney University Stellar Interferometer (SUSI) instrument, currently working at a baseline of 160 metres, is eventually planned to operate at 640 metres. This will ultimately give a resolution of 75 microarcseconds ($0''.000075$) and will allow binaries with periods of hours to be observed directly.

Peter Lawson's website[6] covers all the current interferometer projects and has links to the historical ones.

Two planned satellites, *DIVA* and *GAIA*, will certainly make a significant contribution to our knowledge of binary stars. *DIVA* (Double Interferometer for Visual Interferometry) is a small Fizeau interferometer planned to be mounted on an Earth-orbiting satellite and it is planned to fly in the new few years. If it flies, and at the time of writing this is highly uncertain, it will operate with a scanning law similar to *Hipparcos* and carry out an astrometric and photometric survey down to $V = 15$. High-precision positions, parallaxes, proper motion and photometry will be done for 35 million stars. With a resolution of about 0.5″, and a spectral capability it is expected that this survey will reveal several hundred thousand new binaries by either direct resolution, astrometric shift, anomalous spectral signals or eclipsing systems detected from the photometry.

GAIA is not due to fly until about 2010, but it is estimated that tens of millions of new double stars will de detected. For the resolved pairs, the magnitude difference is important. Equally bright pairs (<15th magnitude) will probably be completely resolved at 10 mas, while a 20th magnitude companion would be seen only at some 50 mas. Closer pairs will be observed by their photocentres but, in the "favourable" period range 1–10 years, a large proportion of them will have their astrometric orbits determined. This will be possible for photocentric orbit sizes below 1 mas, at least for the brighter systems. Bright (again <15th magnitude) shorter-period systems (days/months) will be observed by the radial-velocity instrument (at 0.1 mas separation), and millions of (mainly even shorter-period) eclipsing binaries will be observed photometrically.

References

1   Toomer, G.J., 1984, Ptolemy's Almagest, Duckworth.
2   Hartkopf, W. I. and Mason, B.D., 2001, http://ad.usno.navy.mil/wds/orb6.html
3   Tokovinin, A.A., 1997, Astron. Astrophys. Suppl., **124**, 75.
4   Original article by U. Fedele, 1949, Coelum, 17, 65, but see the web page of Leos Ondra at http://leo.astronomy.cz/mizar/article.htm
5   Pease, F.G., 1927, Publ. Astron. Soc. Pacific, 39, 313.
6   Lawson, Peter http://olbin.jpl.nasa.gov/links

Chapter 2

Why Observe Double Stars?

Bob Argyle

Introduction

Like many branches of astronomy, the observation of double stars can be appreciated at several levels. For those who enjoy the night sky, double stars offer some of the most attractive sights around and they are particularly good in small telescopes where the colours are much more obvious. For a good list of the most impressive pairs, consult the list of 100 best pairs on the Astronomical League Double Star Club website[1] or lists of pairs in *Sky & Telescope* and other journals.[2-8]

Some observers use double stars as a test object to see what their telescope is capable of in terms of angular resolution. Tables 2.1–2.6 give a range of test pairs for both binoculars and telescopes with a range of apertures from 9 cm to 60 cm.

A few observers find double stars to be so endlessly fascinating that they wish to make useful contributions to the subject. This may be by making measures of ρ and θ for the binary systems using a micrometer, doing photometry of wider pairs with a CCD camera or calculating orbits from the observed positions. Most of this book will be dedicated to the description of such techniques and opportunities for useful work are discussed further in Chapter 19.

Colours

Much has been written on this subject and it will continue to exercise fascination amongst observers. It is perhaps the most compelling reason why people observe double stars. Although watching the stars swing around their huge orbits over the years can also be interesting, it does not strike with the same immediacy.

Here some optical aid makes all the difference. With the naked eye, few colours can be ascertained. The contrast between the reddish-orange Betelgeuse and the white Rigel in Orion can be seen and the deep red of Antares certainly stands out, but none of the more subtle colours visible in telescopes appear. Colours tend to be much easier to see when some optical aid is used, for a number of reasons. Firstly, there is more light incident on the eye and the cones, which are small receptors in the eye which detect colour can be more easily stimulated. Next, if the telescope is then deliberately defocused, the star colours become more prominent. The reason for this appears to be pychological in origin. Thirdly, star colours become more intense when contrasted with other stars of different hues. In some double stars such as iota Cancri the companion (distant 30″) appears blue alongside the orange-yellow of the primary star. Yet the spectral types of G7 and A3 indicate that the secondary star should be white and it is simply the contrast with the primary which gives the star its blue colour. In alpha Herculis, the companion which is less than 5″ away is distinctly green although no single stars of this colour are known to exist. (Some observers have reported that Beta Librae is green or pale green but Robert Burnham who mentions this in his *Handbook*, states that the star is white.) It might be interesting to see how the contrast effect varies as the distance between the two stars in a double star system, for stars of similarly different spectral types and brightnesses.

Whilst a telescope enhances the colours in double stars, if too large an aperture is used as Dennis di Cicco[9] pointed out some years ago, colour perception is made more difficult. This can be partly explained by the fact that a smaller telescope produces a larger diffraction disk and the eye is more susceptible to colour in extended images than in point sources.

Colours can be determined in a more systematic manner than by eye estimates which are affected by

Figure 2.1. A CCD image of Albireo (β Cygni) taken from Australia by Steve Crouch. The separation is 34″.7. North is at the bottom, east to the right.

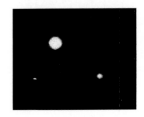

personal equation. One method is to take colour slides of double stars and project the resulting images against a commercially available colour chart (such as the Macbeth Color Checker) to determine the colour of each component. Such a project was carried out some years ago by a group led by Joseph Kaznica and others[10] at the Mount Cuba Observatory in Delaware.

Tests of Resolution

Binocular Tests

Before the appearance of stabilising binoculars it used to be thought that the best resolution available with the standard pair of 7 × 50 binoculars was around 25″. The limiting magnitude also improves with the field being more stable and again it would be most interesting to see what the limit of these instruments is. Table 2.1 lists a number of test objects.

Resolution Tests for Binoculars

Table 2.1 gives a list of 50 double stars that are suitable tests for image-stabilised (and other) binoculars. The pairs have been selected from the WDS with the criteria that the magnitudes should both be brighter than 8.0 and the separations lie between 8 and 25″. The pairs are well distributed around the sky so a number of them will be visible at any time of year. The positions are given for J2000 and the position angle and separation (in arcseconds and degrees respectively) refer to the date given in the previous column. In most cases the motion is very small but a number of these pairs are binary and are indicated by an asterisk (*) after the catalogue name. The magnitudes are visual and come from the WDS. The components AB refer to the brightest two stars in a multiple system. If no components are stated, this means that the given pair is a double only. For an explanation of the catalogue names, see Chapter 24.

Resolution Tests for Telescopes

Tables 2.2–2.7 present some tests of resolution for telescopes of apertures ranging from 90 mm to 60 cm.

Observing and Measuring Visual Double Stars

Table 2.1. Resolution tests for binoculars.

Catalogue	Component	RA(2000)Dec	Date	PA	Separation	V_a	V_b
STF3053	AB	00026+6606	1992	72	15.1	5.96	7.17
STF60*	AB	00491+5749	1999	318	12.7	3.52	7.36
STF100	AB	01137+0735	1994	63	23.2	5.22	6.15
H58		01590–2255	1991	303	8.5	7.28	7.56
STF205	A–BC	02039+4220	1995	64	9.8	2.31	5.02
STF239		02174+2845	1997	212	14.0	7.09	7.83
PZ2		02583–4018	1997	90	8.3	3.20	4.12
STF401		03313+2734	1998	270	11.5	6.58	6.93
STF550	AB	04320+5355	1996	309	10.5	5.78	6.82
STF590		04436–0848	1991	318	9.3	6.74	6.78
STF630	A–B	05020+0137	1997	50	14.4	6.5	7.71
STF688		05193–1045	1991	95	10.6	7.52	7.55
STF872	AB	06156+3609	1991	216	11.4	6.89	7.38
DUN30	AB–C	06298–5014	1994	313	12.0	5.97	7.98
HWE 13		06358–1606	1991	117	12.6	7.37	7.39
STF948	AC	06462+5927	1998	310	8.9	5.44	7.05
STF1044		07164+4738	1991	168	12.6	7.70	7.72
STF1065		07223+5009	1991	255	14.9	7.51	7.67
19		07343–2328	1991	117	9.8	5.82	5.85
STF1122		07459+6509	1991	186	14.9	7.78	7.80
STF1245	AB	08358+0637	1991	26	10.2	5.98	7.16
STF1315		09128+6141	1991	27	24.8	7.33	7.65
SHJ110	AC	10040–1806	1991	274	21.2	6.22	6.97
STT219		10302+5100	1991	300	14.1	7.56	7.70
DUN97	AB	10432–6110	1991	174	12.3	6.59	7.88
BSO6		11286–4240	1991	169	12.9	5.13	7.38
DUN117	AB	12048–6200	1994	150	22.7	7.40	7.83
STF1627		12182–0357	1993	196	19.8	6.55	6.90
STF1694		12492+8325	1997	329	21.5	5.29	5.74
STF1744	AB	13239+5456	1999	153	14.6	2.23	3.88
STF1821		14135+5147	1999	236	13.7	4.53	6.62
HJ4690	Aa–B	14373–4608	1991	25	19.3	5.55	7.65
STF1919		15127+1917	1995	11	23.4	6.71	7.38
LAL123	A–BC	15332–2429	1991	301	9.1	6.94	7.00
PZ4		15569–3358	1991	49	10.3	5.09	5.56
H7	AC	16054–1948	1991	20	13.6	2.59	4.52
DUN206	AC	16413–4846	1991	265	9.6	5.71	6.76
STF2202	AB	17446+0235	1994	93	21.0	6.13	6.47
STF2273	AB	17592+6409	1999	283	21.3	7.31	7.63
SHJ 264	AB–C	18187–1837	1991	51	17.3	6.86	7.63
STF2417	AB	18562+0412	1997	104	23.1	4.59	4.93
STF2474	Aa–B	19091+3436	1997	264	16.1	6.78	7.88
STF2578	AB	19457+3605	1995	125	15.0	6.37	7.04
SHJ324		20299–1835	2000	238	21.9	5.91	6.68
STF2727*		20467+1607	2000	266	9.2	4.36	5.03

Table 2.1. Resolution tests for binoculars. (*continued*)

Catalogue	Component	RA(2000)Dec	Date	PA	Separation	V_a	V_b
STF2769		21105+2227	1999	299	18.0	6.65	7.42
STF2840	AB	21520+5548	2000	195	18.0	5.64	6.42
STF2873	AB	21582+8252	1996	69	13.6	7.00	7.47
DUN246		23072–5041	1995	255	8.8	6.29	7.05

These pairs are chosen because they appear to be moving fairly slowly at the present time and the following list should be accurate until about 2005. The pairs are chosen from the CHARA 4th catalogue of interferometric measures[11] and the values given below are for the epoch 2002.0. The complete catalogue is available on the CD-ROM.

The closest pair in each list corresponds approximately to the Dawes limit for that aperture ($11''.6/D$ in cm) although the magnitude of both components varies so that the fainter and more unequal pairs will be more

Table 2.2. Tests for 90-mm aperture (HIP denotes the Hipparcos Catalogue number).

Catalogue	Component	RA(2000)Dec	HIP	PA	Separation	V_a	V_b
BU728		23522+4331	17655	8	1.20	8.04	8.32
STF367		03140+0044	5058	134	1.13	7.36	7.41
A 1606		13128+4030	64464	15	1.25	8.81	8.91
BU114		13343–0837		168	1.30	(8.05	8.1)
STF987		06541–0552	33154	177	1.30	6.39	6.55
STF2843	AB	21516+6545	107893	147	1.40	6.37	6.67
STF2583		19487+1148	97473	107	1.35	5.75	6.04
STF1291		08542+3034	43721	312	1.49	5.40	5.62
STF314	AB–C	02529+5304	13424	313	1.54	(6.5	6.7)
STF1932		15183+2649	74893	262	1.60	6.59	6.64
STF1639	AB	12244+2535	60525	324	1.67	6.41	7.55

Table 2.3. Tests for 15-cm aperture.

Catalogue	Component	RA(2000)Dec	HIP	PA	Separation	V_a	V_b
BU341		13038–2035	63738	311	0.75	5.58	5.62
BU316		04528–0517	22692	183	0.85	7.71	7.75
BU232	AB	00504+5038	3926	250	0.85	7.83	8.07
STF13		00163+7653	1296	53	0.90	6.35	6.60
STT403		20143+4206	99749	172	0.94	(7.3	7.5)
HO475	AB	22327+2625		309	1.05	(9.3	9.5)
BU694	AB	22030+4439	108845	3	1.10	5.57	7.66

Table 2.4. Tests for 20-cm aperture.

Catalogue	Component	RA(2000)Dec	HIP	PA	Separation	V_a	V_b
A 1504	AB	00287+3718	2252	41	0.54	8.12	8.22
HU517	AB	01037+5026	4971	29	0.56	8.22	8.27
A 347		14369+4813	11467	252	0.57	7.73	7.93
HO44		10121–0613	49961	204	0.58	7.96	8.27
COU482		14213+3050		122	0.60	(9.2	9.3)
HU149		15246+5413	75425	273	0.60	6.68	6.80
BU303		01096+2348	5444	293	0.62	6.65	6.78
HU146		15210+2104	75117	127	0.66	8.82	9.09
BU991		22136+5234		140	0.66	(8.8	8.8)
STT435		21214+0254	105438	235	0.66	7.41	7.46
I 78		11336–4035	56931	98	0.67	5.39	5.44
A 185		22201+4625		319	0.69	(9.6	9.7)
STF412	AB	03345+2428	16664	356	0.70	5.95	5.98
STF2783		21141+5818	104812	355	0.70	7.11	7.34
STF1555	AB	11363+2747	56601	147	0.71	5.80	6.01
STF3056	AB	00046+3416	374	144	0.72	7.02	7.30
A 1116		15116+1008	74348	51	0.77	7.97	7.99
A 2419		03372+0121		96	0.78	(8.6	8.7)
KUI97		20295+5604	101084	132	0.79	5.89	8.77
BU182	AB	23171–1350	114962	47	0.79	8.16	8.38
A 1		01424–0646	7968	248	0.80	8.05	8.20
A 953		01547+5955		65	0.80	(8.8	8.8)
COU610		15329+3121	76127	200	0.82	4.14	6.55
STT112		05398+3758		49	0.84	(7.92	8.2)

difficult to resolve than the bright equal pairs of similar separation.

Note that these lists are merely suggestions for testing telescope objectives, and test objects should not be selected rigorously from one table. Resolution depends, after all, not only on the collimation and quality of the optics, but the state of the atmosphere. It is most likely that the last word on any attempts to resolve close pairs will be had by the seeing, so attempts should be made when atmospheric conditions are suitable.

References

1   Astronomical League http://www.astroleague.com/al/
2   Mullaney, J. and McCall, W., 1965 Nov., *Sky & Telescope*, The Finest Deep-Sky Objects, 280.
3   Mullaney, J. and McCall, W., 1965 Dec., *Sky & Telescope*, The Finest Deep-Sky Objects II, 356.

Table 2.5. Tests for 30-cm aperture.

Catalogue	Component	RA(2000)Dec	HIP	PA	Separation	V_a	V_b
VOU36		02513+0141		9	0.38	(8.4	8.9)
STT75		04186+6029	20105	181	0.38	7.33	7.49
BU688	AB	21426+4103	107137	197	0.38	7.55	7.61
A 1562		05373+4339	3	52	0.39	(9.0	9.0)
CHR91		20045+4814	98858	211	0.39	6.16	9.64
AC16	AB	19579+2715	98248	232	0.39	7.56	7.77
A 1588		09273−0913	46365	196	0.40	(7.2	7.3)
A 2152	AB	10290+3452	51320	50	0.40	8.52	8.79
RST4534		15089−0610	74116	12	0.41	(8.21	8.2)
RST4220		03038−0542	14255	339	0.42	8.85	8.91
A 2719		06203+0744	30120	65	0.44	6.76	6.83
MCA38	Aa	13100−0532	64238	339	0.44	4.38	6.72
STT349		17530+8354	87534	44	0.45	7.51	7.72
A 951		01512+6021	8629	220	0.45	7.98	8.26
A 914		00366+5608?	2886	26	0.46	7.97	8.05
BU1023		07151+2553	35070	304	0.45	8.34	8.52
A 2016	AB	02287+0840		175	0.46	(9.9	9.9)
YSJ1	Aa	10329−4700	51504	95	0.46	5.02	7.39
BU1184		03483+2223		270	0.46	(8.9	9.1)
BU1298		16595+0942	83143	129	0.46	7.96	8.00
A 1607		13124+5252	64517	14	0.47	9.34	9.43
STT86		04366+1945	21465	4	0.47	7.32	7.34
I 450		01519−2309		222	0.48	(8.6	8.9)
STT337		17505+0715	87325	170	0.48	7.72	7.87
KUI8		02280+0158	11474	38	0.52	6.45	6.66
HU1274		15550−1923	77939	119	0.52	5.95	7.96
COU103		15200+2338		283	0.54	(8.9	8.9)
STT510	AB	23516+4205	117646	304	0.55	7.34	7.41

Table 2.6. Tests for 40-cm aperture.

Catalogue	Component	RA(2000)Dec	HIP	PA	Separation	V_a	V_b
COU452		01510+2551	8600	181	0.29	8.08	9.42
HU981		22306+6138	111112	215	0.29	6.98	7.23
COU1214		01373+4015		175	0.31	(9.6	9.6)
COU1659		01298+4547		26	0.32	(9.0	9.3)
STF346		03055+2515	14376	254	0.34	5.45	5.47
BU1147	AB	23026+4245	113788	352	0.35	5.09	7.26
STT250		12244+4306	60522	349	0.35	7.88	8.02
HU520		01178+4946		166	0.36	(8.09	8.3)
A 1204		20143+3129		143	0.36	(9.4	9.7)
COU1510		02016+4107		131	0.36	(9.6	9.6)
COU2037		05219+3934	25060	143	0.37	7.31	7.54
KR12		01415+6240	7895	291	0.37	7.81	7.88
A 1498		23594+5441	118287	90	0.38	7.73	7.77

Table 2.7. Tests for 60-cm aperture.

Catalogue	Component	RA(2000)Dec	HIP	PA	Separation	V_a	V_b
COU2013		02520+1831		93	0.21	(9.1	9.1)
A 506		06357+2816		36	0.23	(8.6	8.9)
B 2550	AB	01425+5000	7979	277	0.23	8.41	8.58
COU1505		00594+4057	4626	138	0.23	8.55	8.70
HO 98		19081+2705	93994	78	0.24	7.53	7.54
MCA60	Aa–B	20158+2749	99874	147	0.24	4.50	6.65
COU1183		21180+3049	105146	18	0.25	8.13	8.30

4   Mullaney, J. and McCall, W., 1966 Jan., *Sky & Telescope*, The Finest Deep-Sky Objects III, 13.

5   Mitton, J. and MacRobert, A., 1989 Feb., *Sky & Telescope*, Colored Stars, 183

6   Adler, A., 2002 Jan., *Sky & Telescope*, The Season's Prettiest Double Stars, 131.

7   Adler, A., 2002 Jul., *Sky & Telescope*, More Pretty Doubles, 111.

8   Ropelewski, Michael, 1999, *An Atlas of Double Stars*, Webb Society; see http://webbsociety.freeserve.co.uk/notes/doublest01.html

9   Di Cicco, D., 1993, Mar., *Sky & Telescope*, The Delights of Observing Double Stars, 112.

10 Kaznica, J.J. et al., 1984, *Webb Society Double Star Section Circular No 3*.

11 Hartkopf, W.I., Mason, B.D., Wycoff, G.L. and McAlister, H.A., 2002; see http://ad.usno.navy.mil/wds/int4.html

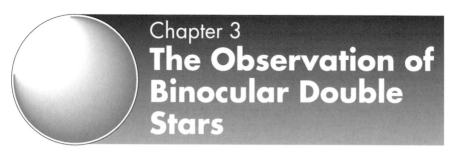

Chapter 3
The Observation of Binocular Double Stars

Mike Ropelewski

Introduction

The night sky presents a fascinating variety of double stars, ranging from wide, optical pairs to close binary systems. A few doubles can be divided with the unaided eye, while a modest pair of binoculars will reveal many more; the study of double stars can be enjoyed by those who do not possess a large telescope or expensive equipment. There is a broad selection of binoculars on the market, so let us take a look at those that might be suitable for this branch of astronomy.

Binocular Features

Probably the best views of celestial objects will be obtained using prismatic binoculars (Figure 3.1). In this design, light passes through the objective lenses and is reflected by prisms before being focused at the eyepieces. Prisms extend the effective focal length of binoculars without increasing their size and create a sharper image than would otherwise be produced. This is especially important when observing double stars; the components should appear as individual pinpoints of light. They also invert the image, resulting in an upright view.

Image stabilised binoculars include advanced design features such as microprocessor controlled variable-angle prisms. These compensate for involuntary move-

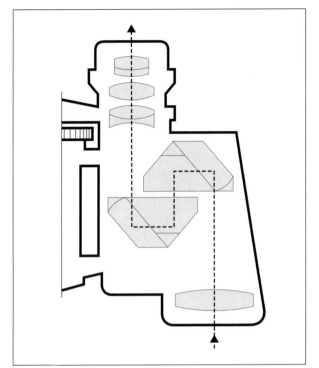

Figure 3.1. The light path in a pair or prismatic binoculars.

ment, enabling the observer to "lock on" to a celestial object at the press of a button. The increased steadiness of the image allows a higher magnification to be used without a tripod or dedicated mount. Comparisons with conventional binoculars have been most impressive. (For a list of test double stars see Chapter 2.)

Another feature of good-quality binoculars is coated or bloomed lenses, where the optical surfaces are treated with a substance to reduce the amount of light reflected from them. The resulting field of view is brighter and free from halos and other false images. Bloomed lenses appear blue or purple when studied under white light – a helpful point to remember if the binocular housing has not been stamped with the words "Coated Optics".

"Optional extras" could include eye-cups, which are circular pieces of plastic or rubber fitted around the eyepieces. Eye-cups prevent stray light from entering the eye and are particularly useful when observing from brightly-lit surroundings.

Most binoculars achieve focusing by means of a manually rotated centre-wheel located on the axis joining the optical systems. Additionally, it is common for one eyepiece to be individually adjustable ensuring that each image is correctly focused for the observer's eyes.

Finally, lens caps. Binocular lenses are delicate items and may incur damage by accidental scratching. A set of tightly fitting covers for eyepieces and objectives will provide valuable protection from mishap and ensure optimum performance is obtained for the lifetime of the binoculars.

Aperture, Magnification and Field Diameter

Having ensured that our binoculars are of decent optical standard, the next points to consider are aperture and magnification. These factors are important because they will determine whether or not a double star can be resolved into two separate sources of light.

Binoculars such as the popular 7 × 50 range (denoting a magnification of seven times and an objective lens diameter of fifty millimetres) are reasonably priced, lightweight and will provide good views of many double stars plus a host of other interesting celestial objects. They are also suitable for general daytime use. Larger instruments with a higher magnification will divide much closer pairs and show greater detail, but are more expensive and bulky. It may be advisable for beginners to invest in a fairly modest pair of binoculars before progressing to an instrument of greater power and aperture, should a deeper interest in astronomy develop.

The field diameter of a pair of binoculars is a numerical value expressed in degrees and fractions of a degree. It is directly related to magnification and objective lens diameter. For a given aperture, field diameter diminishes as magnification increases. As might be expected, it is easier to locate an object through binoculars with a wide field of view, because the area of sky represented is proportionately larger.

To obtain the field diameter of a pair of binoculars, if this value is not known, we need to note the length of time taken for a star near the celestial equator to drift centrally across the field of view from one edge to another (it is necessary to secure the binoculars to a tripod or some other means of support for the test).

Suitable bright stars include δ Orionis (in the belt of Orion), ζ Virginis and α Aquarii. The elapsed time, recorded in minutes and seconds, is multiplied by 15 to give the field diameter in minutes and seconds of arc. This method can also be used for determining the field diameter of a telescope eyepiece.

Binocular Mounts

Conventional hand-held binoculars will resolve the more widely separated double stars, whilst stabilising binoculars, as described above, are capable of dividing much closer, fainter pairs. However some form of mounting is essential if field drawing is to be attempted.

There are several types of adapter. The example illustrated in Figure 3.2 consists of a threaded clamp which is tightened around the central axis of the binoculars; the adapter base is secured to the tripod by means of a standard screw thread. An alternative design comprises an L-shaped bracket with a projecting thread at the top end; this style of adapter is suitable for binoculars that have a threaded recess at the objective end of the central axis. Large binoculars may

Figure 3.2. An example of a simple binocular mount (John Watson).

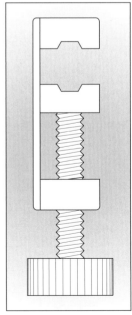
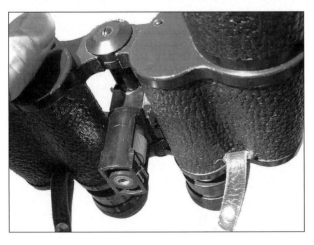

benefit from the extra support provided by the "heavy duty" type of clamp which fits around one side of the binocular housing, giving a more stable and rigid observing platform.

Tripods

It is advisable to choose a tripod that allows binoculars to be secured at slightly above head height, preventing an uncomfortable stoop when studying objects at high altitude. Tripod legs should be strong and sturdy, otherwise any vibration will be transmitted to the field of view, resulting in a shaky image. Both tripod and adapter can be purchased from any good camera shop. Mounted binoculars are portable, easy to set up on any flat, level surface and will enhance the enjoyment of observing double stars and many other celestial features.

What Can We See?

Table 3.1 provides a selection of double stars divisible in binoculars. Positions and measures have been extracted from the Washington Double Star (WDS) catalogue and observational notes have been added by the author. Many of these double stars are marked in *Norton's Star Atlas*[1] which, when supplemented by a publication such as *Sky Catalogue 2000.0*, Vol. 2,[2] will provide both the binocular and telescope observer with a host of interesting objects.

Magnitude and Separation Limits

There are several factors that can affect the magnitude and separation limits (i.e. the faintest stars visible and the minimum separation attainable) for a pair of binoculars. For example, conventional hand-held 7 × 50 binoculars can resolve double stars separated by approximately 1′, whereas image-stabilised binoculars in the 15 × 45 range can typically reduce this to

Table 3.1 Some fine binocular double stars.

RA	2000 Dec	Pair	Comp	Date	PA	Sep	V_a	V_b	Constell.
0149.6	−1041	ENG8	AB	1991	250	184.0	4.66	6.72	Cet
0156.2	+3715	STFA4	AB	1991	298	199.5	5.69	5.89	And
0405.3	+2201	STT559	AB	1993	2	173.9	5.90	7.94	Tau
0433.4	+4304	SHJ44	AB	1993	197	120.3	6.09	6.80	Per
0506.1	+5858	STFA13	AB	1991	9	178.7	5.22	6.08	Cam
0530.2	−4705	DUN21	AD	1991	271	197.7	5.46	6.64	Pic
0535.4	−0525	STFA17	AB	1995	314	134.6	4.90	5.00	Ori
0604.7	−4505	HJ3834	AC	1950	321	196.7	6.00	6.34	Pup
0704.1	+2034	SHJ77	AC	1994	348	100.3	3.80	7.56	Gem
0709.6	+2544	STTA83	AC	1992	80	119.0	7.09	7.77	Gem
0750.9	+3136	FRK7	AB	1991	84	76.8	6.83	7.73	Gem
0814.0	−3619	DUN67	AB	1991	176	66.7	5.09	6.11	Pup
0855.2	−1814	S585	AB	1991	150	64.7	5.75	7.06	Hya
0929.1	−0246	HJ1167	AB	1935	4	65.8	4.60	7.18	Hya
0933.6	−4945	DUN79	AB	1913	32	135.4	7.37	7.50	Vel
1228.9	+2555	STFA21	AB	1991	250	145.2	5.29	6.63	Com
1235.7	−1201	STF1659	AE	1991	275	155.7	7.99	6.70	Crv
1252.2	+1704	STFA23	AB	1991	51	196.1	6.32	6.93	Com
1313.5	+6717	STFA25	AB	1995	296	179.0	6.52	6.96	Dra
1327.1	+6444	STTA123	AB	1994	148	69.6	6.64	7.01	Dra
1350.4	+2117	S656	AB	1993	209	85.5	6.82	7.29	Boo
1416.1	+5643	STF1831	AC	1991	40	109.9	6.68	7.07	UMa
1520.1	+6023	STTA138	AB	1991	196	151.7	7.62	7.74	Dra
1536.0	+3948	STT298	AC	1991	328	122.0	6.78	7.65	Boo
1620.3	−7842	BSO22	AB	1991	10	103.2	4.68	5.27	Aps
1636.2	+5255	STFA30	AC	1994	193	88.9	5.07	5.53	Dra
1732.2	+5511	STFA35	AB	1992	311	61.7	4.86	4.89	Dra.
2013.6	+4644	STFA50	AC	1993	173	107.2	3.80	7.01	Cyg
			AB	1993	323	336.5	3.80	4.80	Cyg
2028.2	+8125	STH7	AC	1991	282	196.8	5.38	6.55	Dra
2037.5	+3134	STFA53	AB	1993	176	181.1	6.32	6.51	Cyg
2110.5	+4742	STTA215	AC	2000	189	135.7	6.40	7.29	Cyg
2113.5	+0713	S781	AB-D	1991	352	183.9	7.15	7.24	Equ
2143.4	+3817	S799	AB	1991	60	150.0	5.69	6.99	Cyg
2144.1	+2845	STF2822	AD	1991	46	198.3	4.49	6.89	Cyg

around 15″ (Sue French, private communication). On the negative side, a bright moon or artificial lighting can create the all too familiar sky-glow that renders faint stars invisible, whilst the presence of atmospheric pollution, cloud or haze can also impair observation. This is most obvious when attempting to study objects located at low altitude; incoming light is more readily absorbed by the thicker layer of atmosphere which may, in severe cases, reduce the apparent brightness of a star by several magnitudes.

Table 3.2. Descriptions of some double and multiple stars from Table 3.1.

Pair	Notes
ENG8	χ Cet. A white and pale yellow double located SW of the orange 4th mag. star ζ Cet.
STFA4	56 And. Pale yellow, pale blue. Lies on the southern border of the open cluster NGC 752.
STT559	39 Tau. Easy white and bluish-white double. East of the yellow 4th mag. star 37 Tau.
SHJ44	57 Per. Superb, bluish-white pair in a field sparkling with many faint stars.
STFA13	11,12 Cam. Bluish-white, pale yellow. Fine pair. A curved chain of four stars following.
DUN21	Orange, blue. Spectacular. Forms a right-angled triangle with two 7th mag. stars.
STFA17	$\theta^1 - \theta^2$ Ori. Two silvery white, 5th mag. twins enveloped by the Orion Nebula.
HJ3834	A neat white pair in a curved E to W chain. The white 4th mag. η Col. lies NW.
SHJ77	ζ Gem. An unequal, yellow and bluish-white couple on a rich background. Tiny comes.
STTA83	A faint, white double in a dense region near the open cluster NGC 2331.
FRK7	A splendid, white pair, 3° E of β Gem. Preceding a dense field.
DUN67	This bluish-white double forms a parallelogram with three other faint stars. Fine area.
S585	A pleasant, pale-yellow pair located south of a W-shaped formation of stars.
HJ1167	τ^1 Hya. White, bluish-white. Unequal. Easily found south of a group resembling Sagitta.
DUN79	An easily resolved pure white couple. The 4th. mag. M Vel. lies N.
STFA21	17 Com. A beautiful, blue pair situated in the Coma Berenices cluster.
STF1659	This white double lies at the NE end of a chain of three tiny stars.
STFA23	32, 33 Com. Pale orange and bluish-white. Lovely contrast. S of the Coma Cluster.
STFA25	A superb orange pair, easily resolved. Situated 3° from STTA123 (see below).
STTA123	Both components are pale yellow. Located in a small arc of fainter stars.
S656	This neat white pair closely follows the yellow 5th mag. star 6 Boo.
STF1831	A splendid, bluish-white double in a field densely populated with tiny stars.
STTA138	This delicate, white pair follows the pale yellow 3rd. mag star ι Dra.
STT298	Both pale yellow. Fine field with 53 Boo (white) and υ Boo (orange) to the NE.
BSO22	A beautiful, golden yellow pair, almost equal in brightness and easy to resolve.
STFA30	Grand, bluish-white pair preceded by a five-star group shaped like a capital X.
STFA35	ν Dra. An exquisite double, comprising two pure white 5th mag. stars.
STFA50	31, o^1 Cyg. Gold, green, blue. A magnificent triple star on the fringes of the Milky Way.
STH7	75 Dra. Both stars orange. A fine, bright pair located in a rich area of sky.
STFA53	48 Cyg. Two pure white "twins" set in a superb region of the Milky Way.
STTA215	Both stars white. Rich area. The orange, 5th mag. 63 Cyg. lies W.
S781	This equal, bluish-white pair is situated near the centre of the Equuleus quadrilateral.
S799	79 Cyg. Both components white. The SE member of a circlet of six stars.
STF2822	μ Cyg. White and bluish-white. Unequal but easy. Set against a rich stellar background.

These "minus points" afflict all visual observers, but should not discourage perusal of the heavens. On a clear, dark night there is much that we can see and do.

Star Colours

A casual look around the sky will reveal that not all the stars are of the same colour. Antares and Betelgeuse, for instance, are orange-red while Altair and Vega appear white or bluish-white. Colour is directly related to a star's surface temperature and the wavelength of the light emitted. Blue or white stars are hotter than those displaying an orange or red hue. Binoculars show the colours well, particularly where the components of a double star present contrasting shades. Examples include θ Tauri, a prominent yellow and white pair in the Hyades cluster, and the superb gold, blue and green triple o^1 Cygni. Conversely, fainter stars on the threshold of visibility appear white because they emit insufficient light to stimulate the colour receptors in the eye.

Occasionally, observers may encounter unusual stellar colours such as violet or mauve. These curious hues are sometimes caused by a phenomenon known as "dazzle tint", where a bright primary imparts false colour to its fainter companion. Star colours are naturally subjective, with opinions often varying between experienced observers. This is just one of the intriguing aspects connected with the study of double stars.

Field Drawing

Perhaps the most enjoyable way of permanently recording a double star observation is to make a field drawing, together with a short written description of the object under study. Sketching trains the eye to notice fine detail and the results can be both personally rewarding and scientifically useful.

Before starting to sketch, it is necessary to prepare some blank circles to represent the field of view. These may be drawn using a pen and template or on a computer/wordprocessor. A field diameter of 6 cm enables six circles to fit on a sheet of A4 paper, allowing sufficient space for captions and notes. For those

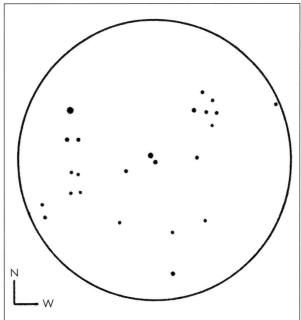

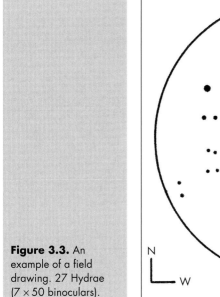

Figure 3.3. An example of a field drawing. 27 Hydrae (7 × 50 binoculars).

N

W

observers who do not own a printer, it may be convenient to produce a page of blank circles which can be photocopied as required.

Other items needed for field drawing are a medium grade black pencil, eraser, sketch-board and a red torch. Especially useful is the "clip-on" design of torch, which can be attached to the drawing board, allowing the observer to sketch more easily.

The next three steps involve finding a light-free observational position, securing the binoculars to a mounting and choosing a suitable double star. Celestial objects near the meridian (due south in the northern hemisphere and due north in the southern hemisphere) are easy to follow because their altitude does not vary much as they cross that part of the sky. After locating the double and before sketching, it may be worth panning the binoculars slightly in altitude and azimuth to obtain the most interesting field of view.

One method of creating a sketch is to begin by drawing the components of the double and the brightest field stars that are visible. Fainter ones can then be added, using the principal stars as reference points. The larger the pencil dot, the brighter the star it represents.

An alternative technique involves dividing the field of view into four equal segments or quadrants and drawing all the stars visible in each section. This approach is, however, probably better suited to telescopic observation, where the field can be accurately divided using an eyepiece fitted with cross-hairs.

The pencil sketch can be overwritten with black ink, if desired, and supplemented by a brief caption. A concise field description could also be included, either with the diagram or, if preferred, in a separate notebook or on a database. An example of a completed field drawing is shown in Figure 3.3. This diagram has been reproduced from the publication *A Visual Atlas of Double Stars*[3] which contains observational details of more than 300 double stars suitable for both binoculars and telescopes.

Summary

The observation of binocular double stars is an absorbing pastime and provides a good introduction to some of the "showpieces" of the night sky. It may lead to more detailed telescopic study of these underrated celestial objects or be enjoyed as a hobby in its own right. Either way, it is a most fascinating branch of astronomy.

References

1   Scagell, R., 1999, *Norton's Star Atlas*, 21st edition.
2   Hirshfield A. and Sinnott, R., 1997, *Sky Catalogue 2000.0*, Vol. 2 (second edition), Cambridge University Press.
3   Ropelewski, Michael, 1999, *A Visual Atlas of Double Stars*, Webb Society (see http://webbsociety.freeserve.co.uk/notes/doublest01.html)

Further Reading

Crossen, C. and Tirion, W., 1992, *Binocular Astronomy*, Willman Bell.

Di Cicco, D., 1998, May, *Sky & Telescope*, Revolutionary New Binoculars, 48.

Moore, P., 1996, *Exploring the Night Sky with Binoculars*, Cambridge University Press.

Peltier, L., 1995, *The Binocular Stargazer*, Kalmbach Publishing.

Seronik, G., July 2000, *Sky & Telescope*, Image-Stabilized Binoculars Aplenty, 59.

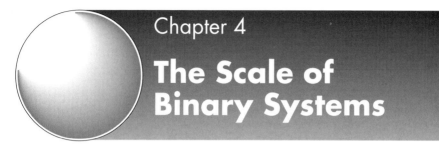

Chapter 4

The Scale of Binary Systems

Bob Argyle

Introduction

How much we can find out about binary systems depends mostly on the separation of the two stars. Very wide pairs with rotation periods of many thousands of years yield little direct information, whilst close pairs with short periods and an orbital plane in the line of sight, thus producing eclipses, will allow many of the individual physical characteristics of the stars, such as mass, size and brightness, to be measured.

The most common type is the visual binary but this is simply due to the fact that these systems are near enough to us that we can resolve them optically. It is quite likely that during the next 10 or 20 years as more sophisticated satellites such as *GAIA* are launched, the number of binary stars known is likely to increase tremendously. This is to be expected since near the Sun we know that more than half our stellar neighbours are members of binary or multiple systems and there is no reason to suppose that this is just a phenomenon peculiar to this region of the Galaxy. At the time of writing the WDS catalogue contains more than 99,000 systems.

Periods Greater than 1000 Years

There is a huge range of scale in binary star orbits and consequently the period can, at the longer end, reach 100,000 years or more. The upper limit is set when the separation of the two stars becomes comparable to the distance to other nearby stars. In this case, the external influences of the neighbourhood stars will eventually disrupt the very tenuous gravitational link between the components of the binary. Periodic passages through the plane of our galaxy (which happens every 30 million years or so) can also disrupt wide binaries due to the influence of giant molecular clouds. It is, of course, impossible to determine these periods even remotely well and even orbital determinations with periods of 1000 years are regarded as very provisional. For the widest systems, the separation of the two stars can reach 10,000 astronomical units (by comparison Pluto is about 30 AU from the Sun and the distance to α Centauri is 280,000 AU).

Periods Between 100 and 1000 Years

Between periods of 100 and 1000 years lie many of the binaries that can be seen with small telescopes such as Castor (445 years), γ Leo (618 years) and γ Vir (169 years). *The Sixth Catalogue of Orbits of Visual Binary Stars*, from the United States Naval Observatory (on the CD-ROM) attempts to list and assess the various orbits which have been calculated for visual binaries. Each orbit is graded from 1 (definitive) to 5 (preliminary) and there are no definitive orbits for binaries with periods greater than that of 70 Oph (88.38 years). (The *Sixth Catalogue* is the regularly updated version available on-line at the USNO website.) This is due to the fact that it is only from around 1830, when F.G.W. Struve was well into his stride at Dorpat working with the 9.6-inch refractor, that reliable (and numerous) measures exist. Clearly it is still important to work on these systems, even though the results may not be used for several centuries. It was the great Danish astronomer Ejnar Hertzsprung who said "If we look back a century or more and ask 'What do we appreciate mostly of the observations made then?' the general answer will be observations bound to time. They can, if missed, never be recovered. Of these observations, measures of double star contribute a major part."

Periods Between 10 and 100 Years

For the serious amateur, pairs with periods between 10 and 100 years are the most rewarding in terms of being able to follow them over a significant portion of their apparent orbit. A good example of a pair in this category is ζ Her with a period of 34.385 years. The apparent separation ranges from 0.5″ to 1.5″, but because the pair is unequally bright (2.8 magnitudes in *V*) when it is near periastron, to see it requires at least a 30-cm aperture. It should be noted that many of these pairs are grade 1 although it is almost certain that *Hipparcos* will have added pairs in this region of which very few observations have ever been made from the ground and which would benefit from further coverage. These are likely to be difficult visually, however. All of the *Hipparcos* discoveries can be found in the WDS catalogue on the CD-ROM. The discovery code is HDS whilst TDS and TDT indicate additional pairs found by the satellite from the Tycho project.

Periods Between 1 and 10 Years

For periods between 1 and 10 years, measures of pairs need large apertures and sometimes special techniques, such as speckle interferometry. Pairs in this region are almost all beyond the range of small apertures.

Periods Less than 1 Year

To detect stars as binary in this period regime, which is beyond the scope of this book, it is necessary to turn to the spectrograph or the ground-based optical array. For an excellent description of the many and varied types of close binary systems see the book by Hilditch.[1]

References

1 Hilditch, R.W., 2001, *An Introduction to Close Binary Stars*, Cambridge University Press.

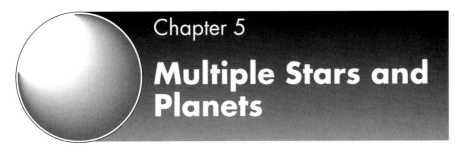

Chapter 5

Multiple Stars and Planets

Bob Argyle

Binary Star Formation

Observational evidence strongly suggests that double stars are the rule rather than the exception in our Galaxy. Recent studies of molecular clouds, using sensitive infrared and millimetre wave detectors (because the visual absorption can exceed 1000 magnitudes), have shown that many of the objects found in these clouds are double or multiple.

Stars are born in dense clouds which consist almost totally of molecular hydrogen along with a small admixture of dust. At the temperature typical of these clouds, about 10 K, the hydrogen cannot be detected. Most clouds also contain traces of carbon monoxide which produces very bright spectral lines at wavelengths of 1.3 and 2.6 mm and it is these which allow astronomers to trace the distribution of hydrogen. To date about 120 other molecules have been found, ranging from water and ammonia to more complex organic structures such as methanol and ethanol.

Molecular clouds come in a range of sizes and composition. The small cloud complex Chamaeleon III, for instance is about 10 pc in diameter, has a maximum visual extinction of a few magnitudes and a temperature of about 10 K. There are a few stars, none of which are massive and no star clusters. The largest complexes in Orion, however, are perhaps 50 pc across, with 100 magnitudes of visual extinction and a gas temperature of 20 K. These are populated by thousands of stars in dense clusters, including massive OB stars. Star formation occurs most frequently in the more massive clouds. Other well-known regions of star formation are known simply by the constellation in which they appear: Taurus – Auriga, Ophiuchus, Lupus, and Perseus, for example.

How then do binary stars form from the nascent interstellar material? Recent simulations on powerful computers can explain not only many of the observed properties of binary stars but also the existence of large numbers of brown dwarfs. These are objects which, in terms of their mass, lie between the massive Jupiter-like planets and the faintest of stars – the red dwarfs. The mass of brown dwarfs (about 0.07 times that of the Sun or alternatively 70 Jupiter masses) is not sufficient for the nuclear reactions in the core to start but they are warm enough to be seen in sensitive infrared detectors.

Bate et al.[1] have recently published the results of collapsing a simulated interstellar cloud in the computer and following its evolution. They begin with a cloud of 50 solar masses and about a light year in diameter and the process starts with the formation of cores which then collapse gravitationally, some being more massive than others. The dense cores are usually surrounded by a dusty disk which is left behind as they contract more and more rapidly. These disks are thought to be the major source for the formation of brown dwarfs. Many interactions occur within the cloud before the stars have reached their full size and as a result the less massive fragments are ejected from the cluster by a slingshot mechanism. The most massive cores are attracted to each other and form close binaries and multiple systems which then undergo further evolution.

When the calculation was stopped (it took 100,000 CPU hours!) the result was the formation of 23 stars and 18 brown dwarfs, so Bate and colleagues conclude that brown dwarfs should be as common as stars. The number of known brown dwarfs is very small but that is largely due to the fact that they are so difficult to detect. Another prediction of this programme is that brown dwarf binaries do form but they need to be very close in order to survive and the few binary brown dwarfs found so far fit this criterion. It was previously thought that the production of close and wide binaries was a result of different processes but this current theory has the advantage of producing many of the observed properties of multiple stars and brown dwarfs.

Planets in Binary Systems

There are two common ways in which planetary bodies (exoplanets) can exist in binary star systems (see Figure 5.1). Firstly, the planet orbits well outside a pair of stars in a close binary orbit. This is referred to as a P-type (or planetary type) orbit. In this case there exists a critical value of the semimajor axis of the planet's orbit around the pair. Too close and the planet is subject to competing pulls from both stars; too distant and the gravitational link vanishes.

Secondly, the planet orbits one or other of a wide pair of stars where the distance of the planet from its sun is much less that the stellar separation. This is an S-type (or satellite-type) orbit and here the semimajor axis of the planetary orbit must be less that a certain critical value if the perturbations from the second star are not to be too disruptive. In other words if the planet wanders too far from its sun during its orbital revolution it will come under the influence of the companion star. All known exoplanets have S-type orbits.

Figure 5.1. Location of stable planetary orbits: **a** the P-type (planetary-type) and **b** the S-type (satellite-type).

At the time of writing, out of the 100 or so planets discovered so far, 17 are known to orbit stars in 13 binary and triple systems. In almost every case planets have been discovered by the reflected variation in radial velocity of the primary star but a recent observation of the star GJ 876 by the *Hubble Space Telescope*

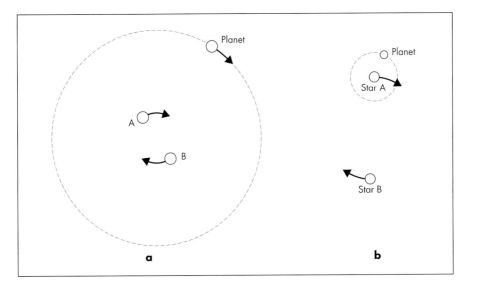

a   b

has revealed the astrometric "wobble" of the primary star to amount to only 0.5 mas caused by the more massive of the two known planets in the system. All the planetary orbits known to date are S-type and are listed in Table 5.1 below. The $M \sin i$ column lists the minimum mass (in Jupiter masses) that the planet has, and the $\sin i$ term represents the unknown inclination of the planetary orbit. Only in one case known to date does an exoplanet eclipse the parent sun giving $\sin i = 1$ ($i = 90°$), so the true planetary mass equals the minimum mass.

The first discovery was a planetary companion to one of the stars in the wide pair 16 Cyg. The planet was detected orbiting the fainter of the two stars which themselves are separated by some 39″ on the sky, equivalent to a linear separation of 700 astronomical units at the distance of 70 light years. The orbital period is very long and nothing is known about the orbit of the two stars about the centre of gravity. 16 Cyg B is a dwarf star, somewhat earlier in spectral type than the Sun. The planet orbits star B at a distance of about 1.72 AU with a period of 800 days but the orbit is very eccentric (0.63). The recent discovery of a very faint star close to A, which is probably physical, means this is the first triple star known to have a planetary companion

55, ρ Cnc is accompanied by a distant M dwarf star which was first identified by W.J. Luyten. The stars make up the system LDS6219. Currently the separation is about 83″ and has shown little change since 1960. The primary star has an annual proper motion of about 0.5″ so it is clearly a physical pair but the orbital period is going to be of the order of thousands of years. Two further planets were confirmed in summer 2002, one of which has the smallest value of $M \sin i$ yet found (0.22).

τ Boo has a faint (magnitude 11.1) M2 companion which was discovered by Otto Struve at Pulkova. At that time (1831) the separation of 15″ was such that the pair could be relatively easily seen. The distance has closed significantly and the current value is around 3″. An orbit was computed in 1998 by A. Hale and a period of 2000 years derived. This is very uncertain but the determination of the binary star orbital elements is significant because from these observations the inclination of the orbit can be determined. If we assume that the planetary orbit around τ Boo is coplanar with that of the two stars then a direct measure of the star's orbital inclination will allow the mass of the planet to be determined directly. If the binary orbit inclination is correct and the tilt of the planetary orbit to the line of sight is also 50° then the sin i factor is 0.77, giving a value of 3.0 M_J for the planet in this system.

The brightest component of the pair STF1341, HD80606, is now known to have a planetary companion with a period of 111.8 days. The eccentricity of the orbit (0.927) is the highest yet found and it is possible that this is due, like that of the planet of 16 Cyg B, to perturbations by the second star in the system.

The wide pair STF2474 consists of two 8th magnitude stars separated by 16″. McAlister found the primary to be a close pair with a period of 3.55 years and recently Zucker et al. found a planetary mass companion to star B which is a G8 dwarf star of 0.87 solar mass.

The bright star γ Cep is a spectroscopic binary of very long period – in fact the longest yet found. Roger Griffin[2] gives the period as 66 years with an uncertainty of 1 year. The planet has a period of 903 days and its average distance from star A is 2.1 AU.

The first planetary discovery made by Italian astronomers with the 3.5-metre Telescopio Nazionale Galileo on La Palma is a low-mass planet orbiting the fainter component of the pair STF 2995 – currently separated by 5.2″. The large proper motion of the bright component and the small change in separation since 1820 confirm that the stellar pair is a binary one.

Table 5.1 summarises the data that we have at present for the binary systems which have planets. The first column gives the popular name of the binary component with the planet, followed by the double star catalogue name, the approximate separation of the two stars (in astronomical units), a letter representing the planet (b = nearest the star, c is next most distant, and so on), and finally the minimum mass of the planet (in terms of the mass of Jupiter). If it were possible to observe the planet by direct imaging, we could determine the inclination of the planet's orbit and hence its mass. If the orbital plane of the planet is in the line of sight then sin i = 1 and the mass of the planet can be determined exactly. This is the case in only one out of the 100 or so planetary systems found to date.

A recent paper by Lowrance et al.[3] lists 11 binary and triple systems which have a planetary companion or planetary system in orbit around one of the stars. Recent discoveries include two more planets in the

Table 5.1. Planets in known double-star systems (January 2003).

Pair	Alias	Separation (AU)	Planet	M sin i (M$_J$)	Notes
16 Cyg B	STF I 46B	700	c	1.5	Triple with 16 Cyg A and a
55, ρ Cnc	LDS6219A	1150	b	0.84	
			c	0.21	
			d	4.05	
τ Boo	STT 270 A	240	b	3.87	
HD 80606	STF1341A	2000	b	3.9	
GJ 86		18	b	4?	
HD179811	STF 2474B	640	b	6.3	Triple with STF 2474 A and a
94 Cet	HJ 663 A	630	b	1.66	
HD 142	HDO 180A	440	b	1.36	
HD 195019	HO 131 A	130	b	3.55	
υ And		750	b	0.68	Optical companion?
			c	1.94	
			d	4.02	
HD 89744		2500	b	7.17	Companion is a brown dwarf
γ Cep	HD 222404	12–32	b	1.76	Planet orbits primary star
HD 219542	STF 2995B		b	0.46	Stars form physical pair
HD114762		120	b		B is a late M dwarf
HD 3651	STT 550		b	0.20	A is nearby (11 l.y.) and K0V

55 Cnc system, a new stellar component to υ And which already has three planets, a faint stellar companion to HD 114762 and a sub-Saturnian mass planet to HD 3651 whose faint stellar companion is a field star.

The website maintained by Jean Schneider[4] at Paris Observatory is kept up to date with new planet discoveries.

Planet discovery is proceeding apace and many further examples are bound to be found in the near future when the upcoming space interferometer missions such as *SIM* and *DARWIN*, which are designed to seek out Earth-sized planets, start operation. We will soon know whether such planets exist in double or even multiple star systems.

References

1 Bate, M. et al. http://www.ukaff.ac.uk/pressreleases/release3.shtml

2   Griffin, R.F., 2002, *Observatory*, **122**, 10
3   Lowrance, P.J., Kirkpatrick, J.D. and Beichman, C.A., 2002, *Astrophys. J.*, **572**, L79.
4   Schneider, Jean http://www.obspm.fr/encycl/encycl.html

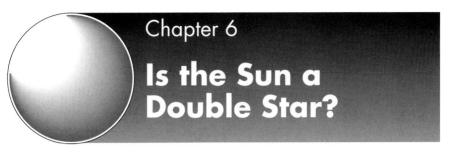

Chapter 6

Is the Sun a Double Star?

Bob Argyle

The Solar Neighbourhood

Within 25 parsecs of the Sun, the number of known stars has risen to almost 3000 but it is still not clear how many more there are to find. Recent estimates indicate that we have only found about half of the stars within this volume of space. The reason for this is that the remaining stars are very faint, but continually improving methods and instrumentation will continue to uncover new neighbours to the Sun in the next few years.

The data in Tables 6.1 and 6.2 is slightly adapted from the *Catalogue of Nearby Stars* courtesy of Dr H. Jahreiss and the Astronomische-Rechen Institut, Heidelberg (June 2002).

The proportion of binary stars in this population can be seen in Table 6.3 where the totals given are accumulating values. Within 5 parsecs half of the stars are in binary systems and another 5% are in triples. However this volume of space represents less than 1% of the volume within 25 parsecs. The average number of stars decreases from 0.118 per cubic parsec within 5 pc to 0.039 per cubic parsec from 20 to 25 pc.

Planetary companions are listed in the penultimate column and at the time of writing the nearest extrasolar planet to the Sun orbits the M dwarf star Gl 876 some 4.6 parsec distant.

Table 6.1. Double and multiple stars within 5 parsecs of the Sun.

CNS	Comp.	RA 2000 Dec	Spectrum	V mag.	B–V	Dist. (l. y.)	Luminosity Sun = 1)
Sun			G2 V	–26.75	0.65	1.0	
Gl 551		14 29 43 –62 40.8	M5.5	11.04	1.81	4.24	0.00006
Gl 559	A	14 39 36 –60 50.1	G2V	0.01	0.71	4.37	1.50
Gl 559	B	14 39 36 –60 50.1	K0V	1.35	0.90	4.37	0.44
Gl 411		11 03 21 +35 58.2	M2V	7.49	1.51	8.31	0.0056
Gl 244	A	06 45 09 –16 43.0	A1V	–1.44	0.00	8.60	22.2
Gl 244	B	06 45 09 –16 43.0	DA2	8.44	–0.03	8.60	0.0025
Gl 65	A	01 39 01 –17 57.0	M5.5V	12.55	1.87	8.73	0.00006
Gl 65	B	01 39 01 –17 57.0	M6V	13.00		8.73	0.00004
Gl 144		03 32 55 –09 27.5	K2V	3.72	0.88	10.50	0.28
Gl 866	A	22 38 33 –15 18.1	M5V	12.87	1.99	11.27	0.00007
Gl 866	B	22 38 33 –15 18.1		13.27		11.27	0.00005
Gl 280	A	07 39 18 +05 13.5	F5IV-V	0.35	0.42	11.41	7.5
Gl 280	B	07 39 18 +05 13.5	DA	10.75		11.41	0.0005
Gl 820	A	21 06 55 +38 44.8	K5V	5.22	1.17	11.41	0.085
Gl 820	B	21 06 55 +38 44.8	K7V	6.04	1.36	11.41	0.040
Gl 725	A	18 42 45 +59 37.9	M3V	8.90	1.53	11.60	0.003
Gl 725	B	18 42 46 +59 37.6	M3.5V	9.69	1.59	11.60	0.0014
Gl 15	A	00 18 23 +44 01.4	M1.5V	8.08	1.56	11.64	0.006
Gl 15	B	00 18 26 +44 01.7	M3.5V	11.05	1.81	11.64	0.0004
Gl 860	A	22 28 00 +57 41.8	M3V	9.79	1.65	13.18	0.002
Gl 860	B	22 28 00 +57 41.8	M4V	11.46	1.8	13.18	0.0004
Gl 234	A	06 29 24 –02 48.8	M4.0V	11.15	1.72	13.43	0.00050
Gl 234	B	06 29 24 –02 48.8	M5.5V	14.24		13.43	0.00003
Gl 473	A	12 33 17 +09 01.3	M5.5V	13.20	1.85	14.31	0.00009
Gl 473	B	12 33 17 +09 01.3	M7	13.19		14.31	0.00009
Gl 687	AB	17 36 25 +68 20.3	M3V	9.16	1.49	14.77	0.004
GJ 1245	A	19 53 54 +44 24.9	M5.5V	13.47	1.92	14.81	0.00007
GJ 1245	B	19 53 55 +44 24.9	M6V	14.01	1.97	14.81	0.00004
GJ 1245	C	19 53 54 +44 24.9		16.78		14.81	0.000003
Gl 876		22 53 17 –14 15.8	M3.5V	10.16	1.58	15.33	0.0016
Gl 412	A	11 05 29 +43 31.6	M1V	8.76	1.55	15.76	0.006
Gl 412	B	11 05 31 +43 31.3	M5.5V	14.41	2.00	15.76	0.00003
Gl 388		10 19 36 +19 52.2	M3V	9.40	1.54	15.94	0.004
Gl 166	A	04 15 16 –07 39.2	K1Ve	4.42	0.82	16.45	0.37
Gl 166	B	04 15 22 –07 39.5	A4	9.51	0.04	16.45	0.0034
Gl 166	C	04 15 22 –07 39.5	M4.5V	11.20	1.65	16.45	0.0007
Gl 702	A	18 05 28 +02 30.0	K0Ve	4.21	0.78	16.59	0.45
Gl 702	B	18 05 28 +02 30.0	K5Ve	6.01		16.59	0.086

Does the Sun have a Companion?

As we have seen the Sun is, as a single star, apparently in a minority amongst the stars in the local neighbour-

Table 6.2. Notes on double and multiple stars within 5 parsecs of the Sun.

Gleise no	Other names	Notes
Gl 551	Proxima Cen; V645 Cen variable 0.01m; double – P = 80 days	
Gl 559A	α Cen A; a = 17$''$.515, P = 79.920 yr	
Gl 559B	α Cen B;	
Gl 411	Lalande 21185; planetary companion?	
Gl 244A	α CMa; Sirius a = 7$''$.500, P = 50.090 yr	
Gl 244B	α CMa B.	
Gl 65A	BL Cet; V(AB) = 11.99, Δm = 0.45	
Gl 65B UV	Cet; a = 1$''$.95, P = 26.52 yr	
Gl 144	ε Eri; dust ring; planetary companion P = 2502 days	
Gl 866A	EZ Aqr; A is SB with Δm = 0, a = 0$''$.34, P = 2.25 yr	
Gl 866B	The system Gl 866 is SB3	
Gl 280A	α CMi; Procyon, a = 4$''$.271, P = 40.82 yr	
Gl 280B	α CMi B; V = 10.75 (HST)	
Gl 820A	61 Cyg ; V1803 Cyg.	
Gl 820B	a = 24$''$.4, P = 659 yr	
Gl 725A	BD+59°1915; NSV 11288	
Gl 725B	G227-047; a = 13$''$.88, P = 408 yr	
Gl 15A	GX And;	
Gl 15B	GQ And; sep (AB) 39$''$, 60°	
Gl 860A	Kr 60; V(AB) = 9.59, Δm = 1.	
Gl 860B	DO Cep; a = 2$''$.412, P = 44.6 yr	
Gl 234A	Ross 614; V(AB) = 11.09, Δm = 3.09	
Gl 234B	V577 Mon; a = 1$''$.009, P = 16.60 yr	
Gl 473A	Wolf 424; V(AB) = 12.44, Δm = –0.01.	
Gl 473B	FL Vir; a = 0$''$.715, P = 15.643 yr	
Gl 687AB	BD+68° 946; sep 0$''$.307, 1°2 (1984.4)	
GJ 1245A	V1581 Cyg; V(AC) = 13.41, Δm = 3.31	
GJ 1245B	G208-045; sep (AB) 7$''$.969, 98.04	
GJ 1245C	a(AC) = 0$''$.28, P = 15.22 yr	
Gl 876	IL Aqr; 2 planetary companions P = 30.1 and 61.0 days	
Gl 412A	BD +44° 2051;	
Gl 412B	WX UMa; sep(AB) = 28$''$, PA = 133°	
Gl 388 A	D Leo; resolved by speckle (?) Heintz: no companion.	
Gl 166A	o^2 Eri	
Gl 166B	40 Eri; P(BC) = 252.1 yr	
Gl 166C	DY Eri;	
Gl 702A	70 Oph A;	
Gl 702B	70 Oph B; V(AB) = 4.02, Δm = 1, a = 4$''$.545, P = 88.13 yr	

Table 6.3. Distribution of single and multiple stars near the Sun

Distance	Volume	Single	Binaries	Triples	Multiples	Planets	Binary freq.
0–5 pc	0.8%	31	31	3		1	50%
5–10 pc	5.6%	152	53	13	6		45%
10–15 pc	15.2%	417	127	13	1	7	39%
15–20 pc	29.6%	662	172	17	1	8	35%
20–25 pc	48.8%	793	210	25	5	3	36%

hood. As more very faint companions to nearby stars are found this will make it even more unusual, but do we really live in a Solar System with a single Sun?

In 1984 Raup and Sepkowski[1] reported evidence for a 26 million year (Myr) periodicity in the occurrence of mass extinctions based on a study of marine fossils. Such impacts included the one 65 million years ago that produced the Chicxulub crater in Yucatan and killed the dinosaurs. Steel[2] refers to later work by Sepkowski which indicates ten such events over the last 260 million years which strongly correlate with a 26 Myr cycle.

This produced a flurry of interest from astronomers who came up with several ideas on how this could be linked to astronomical events. One idea related to the rotation of the Solar System around the Galaxy. It is well established that one rotation around the Galactic centre takes about 250 Myr but during this time the Sun also moves perpendicular to the Galactic plane in a sinusoidal fashion and crosses the plane every 30 Myr or so, reaching a distance of about 100 pc from the plane at the ends of the cycle. During the plane passage, it is surmised, the Earth's biosphere can be exposed to increased levels of radiation. (A recent theory speculates that another intense source of radiation may emanate from supernovae which tend to occur in the Galactic plane.) Rampino and Stothers in *Nature*[3] argued that the original Rapp and Sepkowski data could be interpreted as having a period of 30 Myr rather than 26, and then stated that this is in better agreement with the periodic Galactic-plane crossing period of 33 Myr. With the Sun spending more than two-thirds of its time within 60 pc of the Galactic plane there was ample opportunity for encounters with passing giant molecular clouds to disturb comets from the Oort cloud. Rampino and Stothers also found a periodic term of 31 Myr in the occurrence of large craters on the Earth.

In the same edition of *Nature* the American astronomers Whitmire and Jackson,[4] and, independently, Davis, Hut and Muller[5] came up with a theory to try and explain the apparent 26 Myr periodicity. Whitmire and Jackson postulated a star with mass between 0.0002 and 0.07 $M_\odot$ ($M_\odot$ is the mass of the Sun), with an orbit of eccentricity 0.9 and semimajor axis of 88,000 AU. The companion postulated by Davis et al. was similar but at apastron such an orbit would take it out to a distance of about 3 light years after which the companion would then approach the Sun, skirt the Oort cloud, disrupting comets into the inner Solar System and return again to the depths of space. This companion star was named Nemesis to reflect the catastrophes that its appearances would trigger. Detractors from the theory argued that when at apastron passing stars would have more effect on Nemesis than the Sun, but work by the Dutch astronomer Piet Hut argued that Nemesis could survive such encounters for about a billion years. Today it is difficult to explain binary orbits on this scale. None out of the thousand or so binary orbits which have been catalogued have aphelion distances on this scale.

The main argument against the Nemesis theory is that the projected orbit is too large and too eccentric to allow the star to stay bound to the Sun after more than a few passages through the Galactic plane

Recent studies of wide binaries[6] conclude that some wide pairs have separations in excess of 10,000 AU. To give an idea of this scale, Pluto is about 30 AU away and α Centauri is about 280,000 AU distant.

If Nemesis exists then clearly it is not a twin of the Sun because even at apastron it would have apparent magnitude +3 and its parallax of well over 1″ would have marked it out many years ago. Nemesis must be at least a faint red dwarf, perhaps even a brown dwarf whose apparent magnitude is likely to be at least +15. The proper motion of such a star will be very small and this will be a distinguishing feature as many very faint nearby stars have large proper motions. So a survey such as the Sloan Digital Survey could pick it up, providing the star lies in the 25% of the sky which the survey will cover. Any suitable candidates could then be observed individually by ground-based telescopes since the parallax will be large.

Could the extinction in the late Eocene period be due to a passing star? One possibility of resolving this question may come with data from the projected *GAIA* mission. The expected accuracy of the proper motion and parallax determination for the stars in the solar neighbourhood will allow a more accurate backward interpolation to determine the history of close stellar approaches to the Solar System.

References

1   Raup, D.M. and Sepkowski, J.J., 1984, *Proc. Natl. Acad. Sci. USA*, **81**, 801.
2   Steel, D., 1995, *Rogue Asteroids and Doomsday Comets*, Wiley.
3   Rampino, M.R. and Stothers, R.B., 1984, *Nature*, **308**, 709.
4   Whitmire, D.P. and Jackson, A.A., 1984, *Nature*, **308**, 713.
5   Davis, M., Hut, P. and Muller, P.A., 1984, *Nature*, **308**, 715
6   Allen, C., Herrera, M.A. and Poveda, A., 1999, *Astrophys. Space Sci.*, **265**, 233.

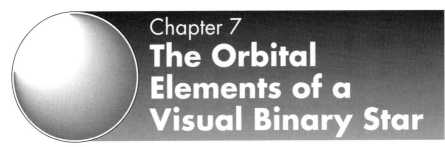

Chapter 7
The Orbital Elements of a Visual Binary Star

Andreas Alzner

The True Orbit

Whilst astronomers regard the brighter component as fixed and map the motion of the fainter one around it, in reality, both stars in a binary system move in ellipses around the common centre of gravity. The size of the ellipse is directly proportional to the mass of the star, so in the Sirius system, for instance, the primary has a mass of 1.5 $M_\odot$, the white dwarf companion 1.0 $M_\odot$ and so the size of ellipses traced out on the sky are in the ratio 1.0 to 1.5 for the primary and secondary (Figure 7.1). The ratio of the masses is inversely proportional to the size of the apparent orbits (see eqn 1.1 in Chapter 1), so this gives one relation between the two masses. To get the sum of the masses requires the determination of the true orbit from the apparent orbit and this is what this chapter will describe.

We regard the primary star as fixed and measure the motion of the secondary star with respect to it, and in Chapter 1 we saw that in binary stars the motion of the secondary star with respect to the primary is an ellipse. This is called the apparent ellipse or orbit and is the projection of the true orbit on the plane of the sky. Since the eccentricities of true orbits can vary from circular to extremely elliptical (in practice the highest eccentricity so far observed is 0.975), then the range of apparent ellipses is even more varied because the true orbit can be tilted in two dimensions at any angle to the line of sight. We need the true orbit in order to determine the sum of the masses of the two stars in the

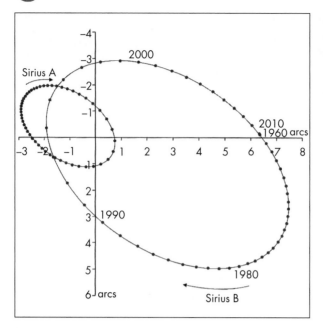

Figure 7.1. The real orbits of the stars in the Sirius system.

binary. This is still the only direct means of finding stellar masses.

On the face of it then the measurements that we make of separation and position angle at a range of epochs are all the information that we have to try and disentangle the true orbit from the apparent orbit. We do, however also know the time at which each observation was made much more precisely than either of the measured quantities. There are other clues, for instance, in the way that the companion moves in the apparent orbit.

In Figure 7.2 I plot the apparent motion of the binary OΣ 363. In this case (x, y) rectangular coordinates are used rather than the θ, ρ polar coordinates which are more familiar to the observer. Each dot on the apparent ellipse represents the position of the companion at two-year intervals. It is immediately clear that the motion is not uniform but it is considerably faster in the third quadrant i.e. between south and west. The point at which the motion is fastest represents the periastron (or closest approach) in *both* the true and the apparent orbits.

Kepler's second law tells us that areas swept out in given times must be equal and this also applies to both the true and the apparent orbit. In Figure 7.2 although the three shaded areas are shown at different points in

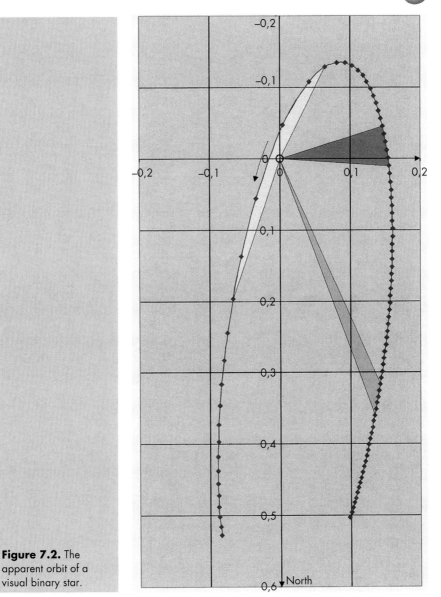

Figure 7.2. The apparent orbit of a visual binary star.

the apparent orbit because they are all traced out over a ten-year interval, the areas are the same. We also know that the centre of the apparent orbit is the projected centre of the true orbit. In most cases the motion is described by the fainter star relative to the brighter star that is fixed in the focus of the ellipse as if the total mass were concentrated in the fixed centre of attraction.

In the true orbit the centre of the ellipse is called C, the focus, and where the brighter star is located is called

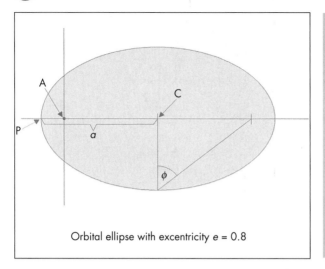

Orbital ellipse with excentricity $e = 0.8$

Figure 7.3. The true elements of a visual binary star.

A. The periastron P is the closest point of the ellipse to A. The geometry of the motion suggests use of polar coordinates. The elements of the real orbit are as follows (Figure 7.3):

P the revolution period in years; alternatively the mean motion
per year ($n = 360/P$ or $\mu = 2\pi/P$ is given);

T the time passage through periastron;

e the numerical eccentricity e of the orbit; the auxiliary angle ϕ is given by $e = \sin \phi$;

a the semiaxis major in arcseconds.

The Apparent Orbit

The apparent (observed) orbit results from a projection of the true orbit onto the celestial sphere (Figure 7.4). Three more elements determine this projection:

Ω the position angle of the ascending node. This is the position angle of the line of intersection between the plane of projection and the true orbital plane. The angle is counted from north to the line of nodes. The ascending node is the node where the motion of the companion is directed away from the Sun. It differs from the second node by 180° and can be determined only by radial-velocity measurements. If the ascending node is unknown, the value < 180° is given.

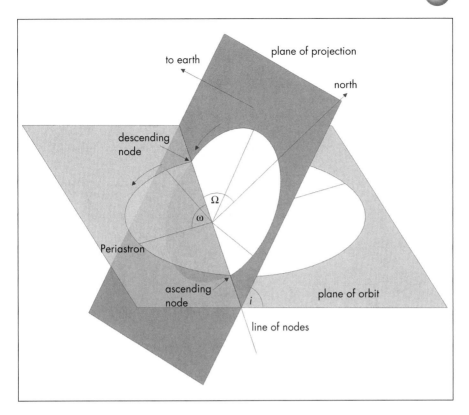

Figure 7.4. The projected elements of a visual binary star.

i   the orbital inclination. This is the angle between the plane of projection and the true orbital plane. Values range from 0° to 180°. For $0° \leq i < 90°$ the motion is called direct. The companion then moves in the direction of increasing position angles (anticlockwise). For $90° < i \leq 180°$ the motion is called retrograde.

ω   the argument of periastron. This is the angle between the node and the periastron, measured in the plane of the true orbit and in the direction of the motion of the companion.

The elements *P*, *T*, *a*, *e*, *i*, ω, Ω, are called the Campbell elements. There is another group of elements which is used in order to calculate rectangular coordinates. They are called Thiele–Innes elements (Figure 7.5):

$$A = a \, (\cos \omega \, \cos \Omega - \sin \omega \, \sin \Omega \, \cos i)$$
$$B = a \, (\cos \omega \, \sin \Omega + \sin \omega \, \cos \Omega \, \cos i)$$

$$F = a \, (-\sin \omega \, \cos \Omega - \cos \omega \, \sin \Omega \, \cos i)$$
$$G = a \, (-\sin \omega \, \sin \Omega + \cos \omega \, \cos \Omega \, \cos i).$$

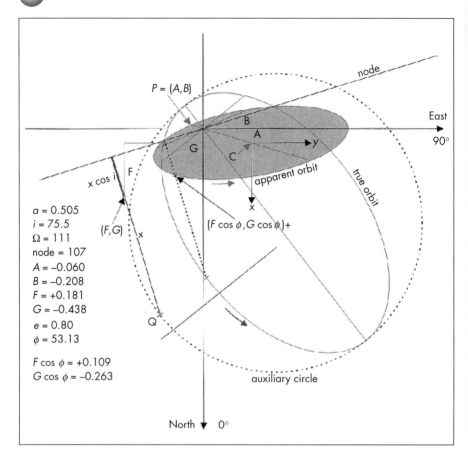

a = 0.505
i = 75.5
Ω = 111
node = 107
A = –0.060
B = –0.208
F = +0.181
G = –0.438
e = 0.80
φ = 53.13

F cos φ = +0.109
G cos φ = –0.263

Figure 7.5.
Thiele–Innes elements and Campbell elements.

Note that the elements A, B, F and G are independent of the eccentricity e. The points (A, B) and (F cos φ, G cos φ), together with the centre of the apparent ellipse, define a pair of conjugate axes which are the projections of the major and minor axes of the true orbit.

There is an instructive and easy way to draw the apparent orbit from the seven Campbell elements. It runs as follows:

1. Draw the rectangular coordinate system with a convenient scale. North is at the bottom (the positive x-axis); east is at the right (the positive y-axis).
2. Draw the line of nodes. The node makes the angle Ω between north and the line of nodes.
3. Lay off the angle ω from the line of nodes and proceeding in the direction of the companion's motion, i.e. clockwise, when i > 90°, and counterclockwise,

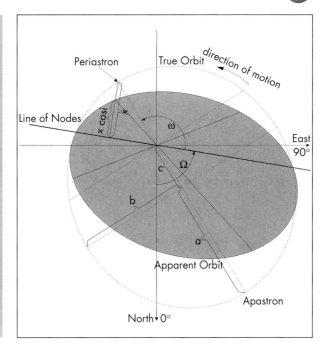

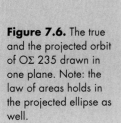

Figure 7.6. The true and the projected orbit of OΣ 235 drawn in one plane. Note: the law of areas holds in the projected ellipse as well.

when $i < 90°$. This will give the line of periastron and apastron of the true orbit.

4. Draw the true orbit ellipse. The distance of the centre of the true orbit from the centre of the coordinate system is c. The long axis is $2a$, the short axis is $2b$, so b and c are easily calculated:

$$c = ae; b = \sqrt{a^2 - c^2}$$

5. Construct the apparent orbit. Draw lines from points on the true orbit to the line of nodes; the lines have to be perpendicular to the line of nodes. Multiply the lines by cos i. Connecting the so obtained points yields the apparent orbit.

As an example, the orbit for OΣ 235 is given in Figure 7.6. Elements are as follows (Heintz[1]): $P = 73.03$ years, $T = 1981.69$, $a = 0''.813$, $e = 0.397$, $i = 47°.3$, $\omega = 130°.9$, $\Omega = 80°.9$.

Ephemeris Formulae

For any time t, the coordinates θ, ρ or x, y are computed from the elements by means of the following formulae. The auxiliary circle has radius a. See Figure 7.7,

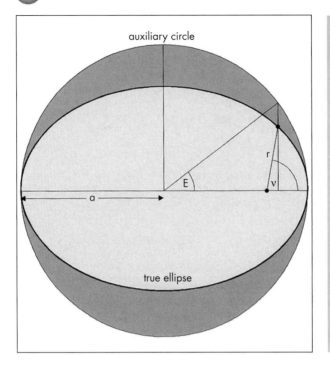
auxiliary circle

true ellipse

Figure 7.7. Auxiliary circle, eccentric anomaly E, true anomaly v and radius vector r.

The angle E is called the eccentric anomaly and has to be determined from the mean anomaly M:

$$\mu\,(t - T) = M = E - \sin E \quad \text{(Kepler's equation)}.$$

This equation is transcendental, i.e. it is not algebraic and has to be solved iteratively. A first approximation is given by the formula:

$$E_0 = M + e \sin M + \frac{e^2}{M} \sin 2M$$

This new E_0 is used to calculate a new M_0:

$$M_0 = E_0 - e \sin E_0$$

A new E_1 is obtained from M, M_0 and E_0:

$$E_1 = E_0 + \frac{M - M_0}{1 - e \cos E_0}$$

The last two formulae are iterated to the desired accuracy. Four iterations are sufficient for $e \leq 0.95$. Now the desired positions are calculated:

Polar coordinates:

$$\tan \frac{v}{2} = \sqrt{\frac{1+e}{1-e}} \, \tan\left(\frac{E}{2}\right)$$

$$r = \frac{a(1-e^2)}{1+e \cos v}$$

$\tan (\theta - \Omega) = \tan (v + \omega) \cos i$

$\rho = r \cos (v + \omega) \sec (\theta - \Omega).$

An alternative formula for the calculation of ρ, due to Michael Greaney,[2] obviates the possibility of the formula becoming undefined, e.g. when $\theta - \Omega = 90°$:

$y = \sin (\theta - \Omega) \cos i$
$x = \cos (\theta - \Omega)$

$$\tan (\theta - \Omega) = \frac{y}{x}$$

$$\rho = r\sqrt{x^2 + y^2}.$$

Rectangular coordinates:

$X = \cos E - e; \quad Y = \sqrt{1-e^2} \sin E$
$x = AX + FY; \quad y = BX + GY.$

References

1   Heintz, W.D., 1990, *Astron. Astrophys. Suppl.*, **82**, 65.
2   Greaney, M.P., 1997, Calculating separation from binary orbits: an alternative expression, *Webb Society Quarterly Journal*, **107**.

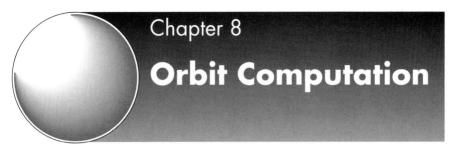

Chapter 8

Orbit Computation

Andreas Alzner

Introduction

Many methods have been given for the calculation of a visual binary orbit. The motion of the Earth can be neglected, but the measurement errors are much larger than errors in positions of planets, asteroids or comets. Therefore these methods are entirely different from calculating an orbit in our planetary system. The decision about whether to calculate an orbit or not may depend on the following considerations:

For the first calculation of an orbit:

- Is the observational material good and complete enough to give a reliable value for the important quantity a^3/P^2?
- Are there only a few recent measurements and does the companion approach a critical phase of the orbit, so that a first preliminary result will attract the observer's attention to the pair?

For the improvement of an orbit:

- Are there large (or growing) deviations between observed positions and calculated positions?
- Will the new orbit give a significantly more reliable result for a^3/P^2?

Rating the observational material: with a strongly marked curvature, even a comparatively short arc may suffice to give a reliable orbit, provided that the observations are consistent; see the two "well-determined"

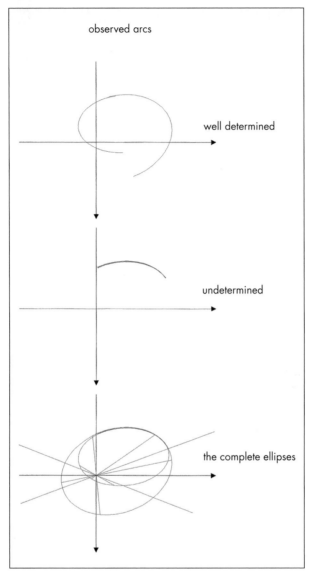

observed arcs

well determined

undetermined

the complete ellipses

Figure 8.1. Well-determined arcs, undetermined arcs and the complete ellipses.

arcs in Figure 8.1. Now have a look at the two "undetermined" arcs. Even high-precision measurements will not allow us to calculate a preliminary orbit. Any result will have to be graded "undetermined". Substantial revisions are to be expected – see the complete ellipses. In the example, the *dotted ellipse* results in a mass seven times larger than the *solid one*!

In the case of the first calculation of an orbit the observed arc will determine which method should be used. If there is any hope that the observational material will allow a least-square fit applied to a set of provisional elements, a simple geometrical method is sufficient to obtain an initial set of elements. If the observed arc is undefined or too short to draw the complete ellipse, a dynamical method is required like the method by Thiele and van den Bos.

Geometrical Methods

The well-observed orbit of $\Sigma 1356 = \omega$ Leo (plot from Mason and Hartkopf[1] is used to illustrate a geometrical method (Figure 8.2). The elements are (van Dessel 1976):

$$P = 118.227 \text{ years}, \ T = 1959.40, \ a = 0\overset{..}{.}880, \ e = 0.557,$$
$$i = 6\overset{.}{.}05, \ \omega = 302\overset{.}{.}65, \ \Omega = 325\overset{.}{.}69 \text{ ascending.}$$

First the apparent orbit is drawn manually. In Figure 8.3 the primary star is located in the centre O of the coordinate system, P is the periastron, A is the apastron, C is the centre of the ellipse, the line connecting the apastron and the periastron is the projected

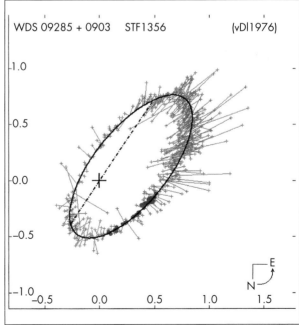

WDS 09285 + 0903   STF1356   (vDl1976)

Figure 8.2. Orbit of $\Sigma 1356$ and observed positions.

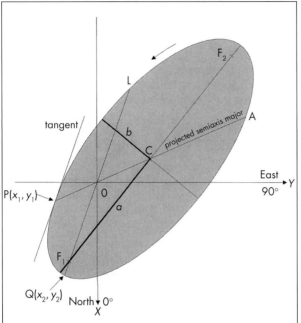

Figure 8.3.
Geometrical method.

semiaxis major, and L and Q are the points where the true anomaly is $-90°$ and $+90°$.

The elements are found as follows:

1. Draw the complete ellipse (the law of areas must be fulfilled). Construct the centre C of the ellipse. After the periastron P is found, the eccentricity is calculated: $e = \text{CO} / \text{CP}$.

2. Draw the tangent in P: first find the focal points F_1 and F_2 of the apparent ellipse. Draw the triangle F_1PF_2. The straight line perpendicular to the line cutting the angle in P into halves is parallel to the tangent in P.

3. Draw the line LOQ: it is parallel to the tangent in P.

4. Determine the coordinates (x, y) of the points $P(x_1, y_1)$ and $Q(x_2, y_2)$.

5. The Thiele–Innes constants are calculated as follows:

$$A = \frac{x_1}{1-e}; B = \frac{y_1}{1-e}$$
$$F = \frac{x_2}{1-e^2}; G = \frac{y_2}{1-e^2}.$$

6. Calculate the elements i, ω and Ω. The relations are:

$$\tan (\Omega + \omega) = \frac{B - F}{A + G}$$

$$\tan (\Omega - \omega) = \frac{B + F}{A - G}$$

$$a^2 (1 + \cos^2 i) = A^2 + B^2 + F^2 \; G^2 = 2u$$

$$a^2 \cos i = AG - BF = v$$

$$a^2 = u + \sqrt{(u + v)(u - v)}.$$

7. Determine the period and the time of periastron from the observed positions. The areal constant c in the apparent ellipse is twice the area of the sector S swept by the projected radius vector in a time Δt:

$$c = 2S / \Delta t.$$

c can be determined graphically from a well observed part of the orbit.

Another definition for the areal constant c is:

$$c = \frac{\mu a^2}{57.296} \cos \phi \cos i.$$

If a and b are the major and minor semiaxes of the apparent ellipse, respectively, the period is:

$$P = \frac{2\pi ab}{|c|}.$$

Now the preliminary elements have to be corrected in a least-square fit. The final result will depend on the weights assigned to the observations. For an easier control of the residuals, it is recommended to separate visual observations, photographic positions and speckle measurements. Given a standard deviation, σ for a measurement, the weight, w, is:

$$w \sim 1 / \sigma^2.$$

The Thiele–van den Bos Method

If the observational material does not allow the entire ellipse to be drawn, the Thiele–van den Bos method is recommended (Figure 8.4). It requires three

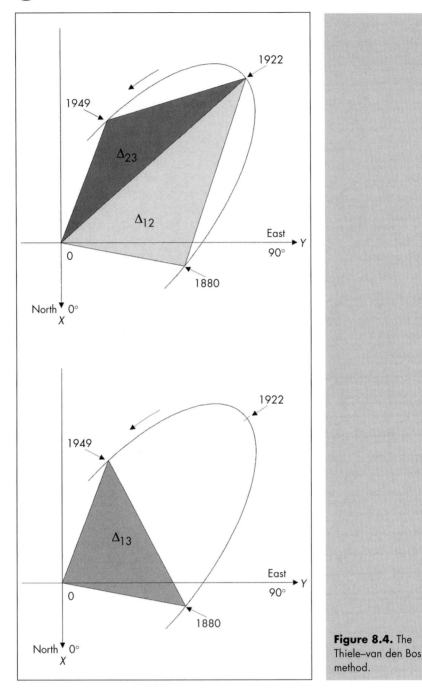

Figure 8.4. The Thiele–van den Bos method.

well-observed places (ρ_i, θ_i) and the corresponding (x_i, y_i) and an approximate value for the areal constant c or alternatively the mean motion μ.

The idea of using the difference

ellipse sector – triangle between three positions

was found by Gauss. For a long time this method was used for the orbit computation of planets, but then Thiele[2] applied the method to binary stars. Van den Bos modified it by inserting the Thiele–Innes constants and was the first to use it in practical work. The following example demonstrates the method. Suppose we are given the following three well-observed positions at three times t_i:

$t_1 = 1880.00$, $\theta_1 = 79.5$, $\rho_1 = 0.546$, $x_1 = +0.099$, $y_1 = +0.537$

$t_2 = 1922.00$, $\theta_2 = 130.9$, $\rho_2 = 1.069$, $x_2 = -0.700$, $y_2 = +0.808$

$t_3 = 1949.00$, $\theta_3 = 158.5$, $\rho_3 = 0.562$, $x_3 = -0.523$, $y_3 = +0.206$.

Note that the three positions have to be determined with the utmost care. If the arc including the three positions does not fulfil the law of areas, no result will be obtained.

The double areal constant c has been determined from the observational material: $c = +0.0139$. Note that in retrograde orbits the sign of c is negative.

For the time interval $t_y - t_x$, Kepler's equation reads as follows:

$$\mu\,(t_y - t_x) = (E_y - E_x) - e\,(\sin E_y - \sin E_x).$$

The double area of the triangles can be calculated from the observed positions:

$$\Delta_{xy} = \rho_x \rho_y \sin\,(\theta_y - \theta_x).$$

For our example we get:

$$\Delta_{12} = 0.4562, \quad \Delta_{23} = 0.2783, \quad \Delta_{13} = 0.3012.$$

The following fundamental relation holds:

$$t_y - t_x - \frac{\Delta_{xy}}{c} = \frac{1}{\mu}[(E_y - E_x) - \sin\,(E_y - E_x)].$$

Introducing the quantities p and q we get:

$$E_2 - E_1 = p; \quad t_2 - t_1 - \Delta_{12}/c = L_{12} \qquad (8.1)$$
$$E_3 - E_2 = q; \quad t_3 - t_2 - \Delta_{23}/c = L_{23} \qquad (8.2)$$
$$E_3 - E_1 = p + q; \quad t_3 - t_1 - \Delta_{13}/c = L_{13} \qquad (8.3)$$
$$\mu\,L_{12} = p - \sin p \qquad (8.4)$$

$$\mu\,L_{23} = q - \sin q \qquad (8.5)$$
$$\mu\,L_{13} = (p + q) - \sin(p + q). \qquad (8.6)$$

First the quantities L_{xy} have to be calculated. Then a μ has to be found by trial and error which satisfies the equations (8.4), (8.5) and (8.6). If the double areal constant c is not known with the required accuracy, but an approximate value for μ is known (for example, because of recurrence of the companion in a position), c has to be found. Our example gives:

$$L_{12} = 9.18; \quad L_{23} = 6.98; \quad L_{13} = 47.33.$$

If the observed arc shows little curvature, i.e. it is undetermined, the values for the L_{xy} become very small and the method will give very uncertain results, if at all. Insertion into the equations (8.4), (8.5) and (8.6) gives the following values for μ, p, q:

$$\mu = 0.0531; \quad p = 1.483; \quad q = 1.344; \quad p + q = 2.827.$$

Now E_2 and e are computed from the following two formulae:

$$e \sin E_2 = \frac{\Delta_{23}\sin p - \Delta_{12}\sin q}{\Delta_{23} + \Delta_{12} - \Delta_{13}}$$
$$e \cos E_2 = \frac{\Delta_{23}\cos p - \Delta_{12}\cos q - \Delta_{13}}{\Delta_{23} + \Delta_{12} - \Delta_{13}}.$$

Another critical point: is the result for e reliable? Up to now no definitely parabolic or hyperbolic orbits have been found!
In our example the result is:

$E_2 = 3.9067; e = 0.557.$

Note that e must be positive.
E_1 and E_3 are obtained from equations (8.1) and (8.2). Equation (8.3) serves as a check. Results:

$$E_1 = 2.4237; E_3 = 5.2507$$

Now the quantities M_x are calculated:

$$M_x = E_x - e \sin E_x.$$

Inserting the three values for M_x into

$$T_x = t_x - \frac{M_x}{\mu}$$

we get three different values for T. The mean value is taken: $T = 1841.17$.

Finally the Thiele–Innes constants are calculated using the relations:

$$x_i = AX_i + FY_i; \quad y_i = BX_i + GY_i$$

a and i are calculated from

$$a^2 (1 + \cos^2 i) = A^2 + B^2 + F^2 + G^2 = 2u$$

$$a^2 \cos i = AG - BF = v$$

$$a^2 = u\sqrt{(u+v)(u-v)}.$$

ω and Ω are calculated from:

$$\tan (\Omega + \omega) = \frac{B - F}{A + G}$$

$$\tan (\Omega - \omega) = \frac{B + F}{A - G}.$$

The quadrant of the node is determined by the relations

$$A + G = 2a \cos (\omega + \Omega) \cos^2 \frac{i}{2}; \quad A - G = 2a \cos (\omega - \Omega) \sin^2 \frac{i}{2}$$

$$B - F = 2a \sin (\omega + \Omega) \cos^2 \frac{i}{2}; \quad -B - F = 2a \sin (\omega - \Omega) \sin^2 \frac{i}{2}$$

In case the ascending node is unknown (i.e. no radial-velocity measurements are available), the value $0 \leq \Omega \leq 180°$ is taken.

For the remaining elements we get:

$$a = 0''.88; \, i = 66°.0 \, ; \Omega = 145°.6; \, \omega = 122°.9.$$

Warning: the Thiele–van den Bos method is instructive, seems elegant and makes full use of the high accuracy of the times t_i but it cannot handle all cases. According to Couteau, it "satisfies the spirit, but not always the investigator".

Now the so-obtained initial elements should be corrected by means of a least-square fit.[3–6] In case the orbit is too uncertain to correct all elements simultaneously one or several elements have to be fixed, limiting the number of elements to be corrected in one step. As a rule, this procedure is an iterative one. The covariance matrix will show which elements are weakly determined or whether there is a strong coupling between two elements. It also allows us to calculate approximate values for the errors of the individual elements, but this result will depend very much on the weights assigned to the observations.

Another way is to define a set of initial values for P, T and e, in the next step calculating the four Thiele–Innes elements by a least-square fit, and varying P, T and e in a three-dimensional grid search.[7]

There are many other methods to get initial elements for a binary star orbit.[8-13] Whatever method is adopted, there is no single method which can handle all cases equally well or which can deliver the final solution in one step.

Edge-on orbits and other special cases have been described in *Double Stars*.[4]

References

1 Mason, B. and Hartkopf, W.I., 1999, *Fifth Orbit Catalogue*, US Naval Observatory.
2 Thiele, T.N., 1883, *Astr. Nachr.*, **104**, 245.
3 Heintz, W.D., 1967, *Acta Astron.*, **17**, 311.
4 Heintz, W.D., 1978, *Double Stars*, Reidel, Dordrecht, Holland.
5 Bos, W.H. van den, 1926, *Union Obs. Circ.*, **2**, 356.
6 Bos, W.H. van den, 1937, *Union Obs. Circ.*, **98**, 337.
7 Hartkopf, W.I., McAlister, H.A. and Franz, O.G., 1989, *Astron. J.*, **98**, 1014.
8 Kowalsky, M., 1873, *Proceedings of the Kasan Imperial University* and 1935 in *The Binary Stars*, by Aitken, R.G., McGraw-Hill.
9 Zwiers, H.J., 1896, *Astr. Nachr.*, **139**, 369.
10 Danjon, A., 1938, *Bull. Astron.*, **11**, 191.
11 Kamp, P. van de, 1947, *Astr. Nachr.*, **52**, 185.
12 Rabe, W., 1951, *Astr. Nachr.*, **280**.
13 Docobo, J.A., 1985, *Celest. Mech.*, **36**, 143.

Chapter 9

Some Famous Double Stars

Bob Argyle

Introduction

In this chapter we move out from the Sun and look at some of the neighbouring double and multiple stars which have been observed for centuries. In some cases there are still secrets to be revealed. The beauty of a sunset on Earth has inspired poets and artists for millennia – what must it be like when there is not one sunset but two or more with each sun glowing in a different colour. The chiaroscuro would be impressive to say the least. Not all double and multiple systems have different colours – some contain stars of essentially the same spectral class and therefore colour.

Mizar and Alcor

This brightest of naked-eye double stars was known in antiquity and attracted the attention of the early telescopic observers. Alcor (magnitude 4.2) is 11.8′ distant, making the pair easy to see. In 1617 Castelli noted that Mizar, the brightest of the two stars ($V = 2.0$) was again double and so Mizar has the distinction of being the first double star discovered at the telescope.

Bradley, in 1755, was the first to measure its relative position at 143°.1, 13″.9. Lewis,[1] using positions up to 1903, found that the annual motion in position angle was +0°.025 and from this he estimated a period of 14,000 years. The physical connection between Mizar and Alcor was established when the proper motions were found to be similar. In fact, there is a stronger connection since a number of the other bright stars in the Plough are moving through space in a loose association – the nearest star cluster to us, in fact twice as close to us as the Hyades. The exceptions to this are α and η.

In 1857 Mizar emerged into prominence once more as it became the first double star to be imaged photographically. Bond used the 15-inch refractor at Harvard College Observatory for this purpose. Agnes Clerke[2] quotes "Double star photography was inaugurated under the auspices of G.P. Bond, Apr 27, 1857 with an impression, obtained in eight seconds, of Mizar, the middle star in the handle of the Plough."

With the advent of photographic spectroscopy, plates of Mizar A taken at Harvard College Observatory in 1886 showed the Calcium K line, leading to an announcement by Pickering in 1889.[3] Mizar A had also become the first spectroscopic binary to be found, beating the discovery of Algol[4] by a few months. In 1906 Frost[5] and Ludendorff[6] independently announced that Mizar B was also a spectroscopic binary, this time a single-lined system of low amplitude, making radial-velocity measurements rather difficult. The period was not determined correctly until relatively recently when Gutmann[7] found a value of 175.5 days.

In the 1920s, with the 20-foot stellar interferometer Frederick Pease[8] carried out two sets of observations, in April 1925 and May/June 1927, calculating a period of 20.53851 days for the orbit of Mizar A (Figure 9.1).

The *Hipparcos* satellite showed that the parallax of Mizar is 41.73 mas whilst that of Alcor is 40.19 mas, corresponding to distances of 23.96 and 24.88 parsecs, thus giving a formal difference in the distance to the two stars of about 3 light years.

In the 1990s the spectroscopic pair Mizar A became one of the first stars to be observed using the Mark III optical interferometer on Mount Wilson in California. An improved instrument, the Navy Prototype Optical Interferometer (NPOI), was then constructed in Arizona. A product of the collaboration between the United States Naval Observatory, the Naval Research Laboratory and Lowell Observatory, the instrument uses phase-closure to build up an optical image of the two components. An interesting consequence of observing pairs with such short periods is that the orbital motion over one night is substantial and has to

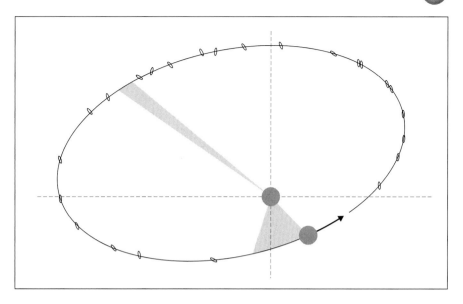

Figure 9.1. The components of the 20.3 day spectroscopic binary Mizar showing motion over a 24-hour period (below). Above is plotted the apparent orbit from NPOI observations. The minimum separation is 4 mas. Note the size of the error ellipse for each observed point. The error ellipse arises because the errors in position angle and separation are not the same magnitude and also depend on the orientation of the interferometer optics with respect to the star being observed. (Courtesy of Dr Christian Hummel, USNO.)

be allowed for. The NPOI data are more accurate than that from the Mark III and allows the dimension of the orbit to be determined without an independent measure of parallax.

The orbit was found to have a semimajor axis of 9.83 mas and the maximum observed separation was 11 mas and the minimum 4 mas. Combined with the data from the spectroscopic orbit, the masses have been determined with great accuracy. The distance has also been derived since both the linear and angular sizes of the orbit are known.

Castor

Possibly found by G.D. Cassini in 1678, the brilliant white leader of Gemini was certainly known to be a double star in 1718 when Bradley and Pound noted the position angle by projecting the line between the two stars and referring it to lines drawn to the nearby bright stars. In 1722 they repeated the observation and a significant change had occurred. Sir John Herschel evaluated this and found that the PA had decreased by more than 7°.

Castor is the pair which Sir William Herschel first used to demonstrate his theory that the motion between the two stars is due to a physical attraction.

In the nineteenth century the large numbers of observations of Castor by double-star observers led to a plethora of orbits with periods ranging from 250 to more than 1000 years. As the pair had not then passed periastron, or even defined one end of the apparent ellipse, this was all preliminary. Even today, several orbits give similar residuals and the period would seem to be of order 450 years. A third star of magnitude 11, Castor C, located at 164° and 71″ (2000) and originally thought to be of use for measuring the parallax of AB is actually moving through space with Castor and is part of the system.

In 1896 Belopolsky showed that Castor B was a single-lined spectroscopic binary whilst Curtis at Lick Observatory[9] showed that the same applied to the A component. In 1920, Adams and Joy[10] announced that Castor C was also a short-period spectroscopic binary but in this case it was double-lined and it also turned out to be an eclipsing system and is now known as YY Gem.

Castor is a relatively nearby system and *Hipparcos* determined a parallax of 63.27 mas equivalent to a distance of 15.80 parsecs or 51.5 light years. From this and the semimajor axis of the orbit one can estimate the real size of the true orbit of Castor AB. The maximum separation of the stars is about 130 AU, some four times the distance of Pluto from the Sun.

Although the bright components A and B are single-lined spectroscopic systems, it was originally assumed that the stars in each system were similar in spectral type. Recent observations of X-ray emission from all three visible stars in the Castor system have proved that the companions to A and B are late-type stars, a conclusion borne out by the distribution of masses in the system. The total mass of the Castor AaBb quadruple is 5.6 $M_\odot$. This is made up of Castor Aa (spectral types A1V and K7V and masses 2.6 and 0.7 $M_\odot$) and Castor Bb (spectral types A1V and M0V and masses 1.7 and 0.6 $M_\odot$). Star C, which is the eclipsing variable YY Gem, is also extremely active in X-ray and radio wavelengths and it is thought that the surfaces of both components are covered in star spots. Its two components are dwarf stars of spectral class M1. A recent paper by Qian[11] speculates that a weak periodic variation in the period of YY Gem may be due to a perturbation by either a brown dwarf or giant planet or it may also be due to magnetic activity, so further research is needed.

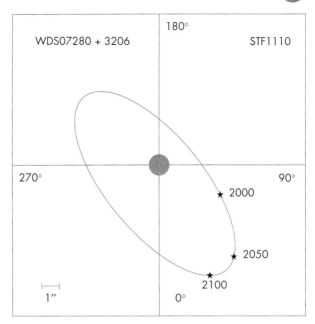

Figure 9.2. The apparent orbit of Castor, period = 445 years. In this and subsequent figures the radius of the central circle represents the Dawes limit for a 20-cm aperture.

Castor, like Mizar, is also part of a moving group that contains 16 other stars including the first magnitude objects Vega and Fomalhaut.

Two current orbits which give small residuals from recent observations show the pair widening for about 80–100 years before it reaches a maximum distance of about 8″ early in the twenty-second century. It will thus remain an easy and beautiful object in small telescopes for many years to come. Figure 9.2 shows the apparent orbit of AB.

xi UMa

This beautiful pair of yellow stars was discovered by William Herschel on May 2 1780, when he wrote "A fine double star, nearly of equal magnitudes, and 2/3 of a diameter asunder; exactly estimated." From the latest orbital elements, we can deduce that the separation on that date was 2.3″. As Herschel was describing the separation between the disks rather than the disk centres we can see that the images in his telescope must have been about 1.4″ across. The subsequent, rapid orbital motion convinced Herschel that the stars were genuinely connected and in 1827 Savary,[12] in France, made

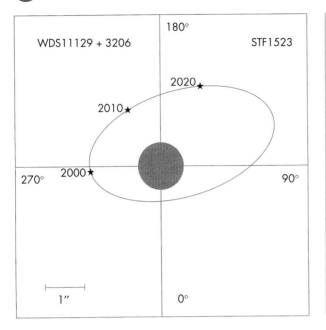

WDS11129 + 3206

180°

STF1523

2020

2010

270°   2000   90°

1″   0°

Figure 9.3. The apparent orbit of xi UMa, period = 59.878 years.

the first orbital analysis of any double star using xi UMa for the purpose. He obtained a period of 58.8 years and an eccentricity of 0.41. This compares with today's latest values of 59.9 years and 0.40. The apparent orbit of xi UMa appears in Figure 9.3.

As was the case with 70 Oph (see following), the ease of measurement of the pair and the relatively short period led to a plethora of orbits. At the beginning of the last century the separation of the pair was over 2″ and increasing, so taking spectra of both components became possible in good seeing. Norlund[13] found a small periodic perturbation in the residuals of the orbit of star A with a period of 1.8 years. As dark companions were somewhat in vogue at the time it seemed natural to ascribe this as the cause of this effect. At this time Wright at Lick Observatory had already noted radial velocity changes corresponding to this 1.8 year period in the spectrum of A. Eventually an orbit was computed by van den Bos[14] which is still used today.

Although spectral plates were also taken of star B from 1902 it was not until 1918 that it, too, was found to be a spectroscopic binary with a period of just under 4 days. Berman[15] produced an orbit for Bb which remained the sole analysis until Griffin revisited the system.[16] He was able to show that Berman's orbit required little adjustment, the difference in the period being 0.6 second! With each successive orbit, the period can be fixed with greater and greater certainty, if the periastron passage is sharply defined. Since Berman's analysis, the pair Bb had gone through more than 6000 orbits.

The next development came much later during an investigation of the system by Mason et al.[17] at CHARA (Georgia State University). By using speckle interferometry measures they were able to obtain very accurate relative positions and these were used in an attempt to tune the orbital elements of the AB pair to give a more precise value of the individual masses (1995). During the course of their observing campaign, Mason et al. observed yet another component, attached to the Bb subsystem but it appeared in only one out of 27 observations.

A later discussion by Daniel Bonneau[18] argues that if this new component exists, it would have a mass of about 0.5–0.7 $M_\odot$ and the orbital inclination of the B system would then be incompatible with both the rotation of B and the coplanarity of the orbit of Bb. Resolution of Bb will only be possible from ground-based interferometer systems although Aa should be resolvable in a 2.5-metre telescope with infrared adaptive optics.

70 Oph

Discovered by William Herschel in 1779, this pair has been a favourite amongst double-star observers of all kinds ever since. Its proximity to the Sun (16.6 light years according to *Hipparcos*) means that during the orbit of 88 years the separation of the stars varies from 1.5 to 6.5″, and it is thus possible to follow it through its whole orbital cycle with ease. The recent periastron passage in 1984 showed the companion moving almost 20° over the year. Another reason for its popularity is the beautiful contrast between its unequal components which have given it a prominent place in all observing handbooks. Placed near the equator it can be seen from virtually all latitudes.

Thomas Lewis in his book on the Struve stars said, in 1906, "It is a splendid system and quite worth the time spent on it by Observers and Computers, although it is a source of much trouble to the latter." Surprisingly enough, it was only recently that the agreement

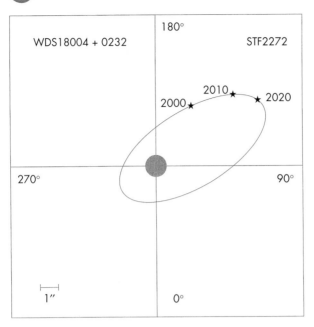

WDS18004 + 0232   STF2272

Figure 9.4. The apparent orbit of 70 Oph, period = 88.38 years.

between the spectroscopic and visual orbit was regarded as satisfactory.

70 Oph was a very popular object with Victorian observers and so measures were numerous. As the pair is an easy object (see the apparent orbit in Figure 9.4) disquiet was expressed about the way that the observed measures were not agreeing with the predicted values from the various and numerous orbits that were being calculated (Lewis lists 22). In 1896, T.J.J. See[19] postulated that these disagreements were due to the presence of a third body orbiting one of the stars in the system. In 1906 Lewis dedicated a large amount of time and space in his volume to discussing the pair. He was convinced that the anomalies were due to a third body orbiting star B and even derived a period of 36 years for it. Burnham, in his catalogue, dismissed the idea saying it was merely observational error but the idea persisted. Pavel[20] postulated a companion orbiting A with a period of 6.5 years.

In 1932 Berman, using radial-velocity measurements of plates taken at Lick Observatory, found a cyclical trend with a period of 18 years but many years later Berman said that he had ceased to be convinced of this result.[21]

Reuyl and Holmberg[22] at McCormick Observatory found an astrometric perturbation with an amplitude of 0.014″ from a series of plates taken between 1914 and 1942.

Worth and Heintz[23] re-visited the visual measures and also produced a trigonometrical parallax for the star. Although there were some problems with measures in the 1870s they could find no evidence for a third body other than a rather unlikely scenario of the passage of a third body through the system at that time.

Heintz computed the orbit afresh in 1988[24] and summarised the situation at the time. This was that recent radial velocity measures showed no perturbation and modern measures using long-focus photography show no systematic deviations beyond the 0.01″ level.

Batten and Fletcher[21] re-examined the radial velocity material measured by Berman and could not find his periodic component in the velocities. However they came to the conclusion that the quality of the early plates means that large residuals "are not of much significance". The re-determination of the spectroscopic period came out at 88.05 years and agrees with Heintz's visual orbit within the quoted error (0.70 year).

It appears that 70 Oph really is what it seems – a beautiful binary star.

Zeta Cnc

The history of this fascinating multiple star has recently been comprehensively reviewed by Roger Griffin[25] but a brief summary is worth including here. Whilst the duplicity of the star had been taken to originate with Tobias Mayer who observed it in 1756, Griffin has shown that the first suspicion that the star was double comes from an observation by John Flamsteed on 22 March 1680. Flamsteed refers to "the north-following component" which agrees nicely with the position of the brighter of the two stars at that time. The separation of the two stars at the time was about 6″.

In November 1781 William Herschel divided the bright component into two and catalogued the close pair as H I 24. (Herschel allocated a class for pairs of differing separations ranging from I for pairs closer than 2″ out to VI for pairs divided by 32″ or more.) His notes on the position angle allow a value of 3.5° to be assigned to the system. The close pair was not observed

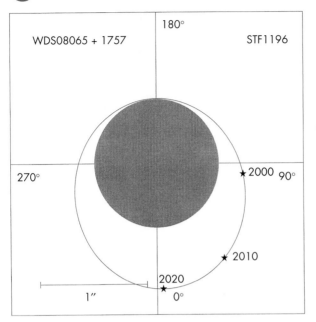

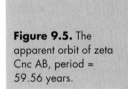

Figure 9.5. The apparent orbit of zeta Cnc AB, period = 59.56 years.

again until 1825 when Sir James South measured it from France when the position angle was given as 58°. It was only when later measures showed that the position angle was actually decreasing that it became clear the close pair had moved through 305° since 1781! The apparent orbit of this beautiful pair is shown in Figure 9.5.

Over the next 20 years or so, growing numbers of double-star observers made copious measures of both the close and wide pair, and the motion of star C around AB was clearly not proceeding in a smooth curve. The position angle would reduce smoothly and then for several years it would stay constant and then resume its course. In 1874 Otto Struve considered the results of almost 50 years of measures by his father, F.G.W. Struve and himself. His conclusion was that the "wobbling" of C was due to the presence of a fourth star D rotating around it with a period of about 20 years. Towards the end of the nineteenth century, Seeliger produced a comprehensive analysis of the motions in the zeta Cnc system. His astrometric orbit for the pair CD remained in force for over 100 years.

Whilst the existence of star D was in no doubt, a few sporadic efforts were made during the last century to detect it. In 1983, D.W. McCarthy [26] using an infrared speckle interferometer announced that he had detected not only star D but yet another component, in other words, the main sequence component C, a white dwarf and another star. This detection was never confirmed and there the matter stood until the early months of 2000.

Using an adaptive optics system working in the infrared on the Canada–France–Hawaii telescope on the island of Hawaii, J.B. Hutchings et al.[27] produced the first direct image of star D. It is a very red object but the effect it has on star C suggests a comparable mass to C, and thus D itself probably comprises a pair of M dwarf stars.

The story does not end here however. In 2000, A. Richichi[28] reported on the observation of a reappearance of zeta Cancri in the 1.52-m telescope at Calar Alto on 7 December 1998. Working in the infrared with a broad-band K filter the occultation trace showed four definite stellar sources and slight but significant evidence for a fifth star, located some 64 mas from star C. Referred to as E, it would appear that it is another low-mass M dwarf possibly with a period of 2 years. The component seen by Hutchings and Griffin, D, was also easily visible but if double the separation is likely to be no more than 30 mas, thus requiring a considerably larger aperture to resolve it.

References

1 Lewis, T., 1906, *Struve's Mensurae Micrometricae*, Mem. RAS, 56.
2 Clerke, Agnes M., 1902, The *History of Astronomy during the Nineteenth Century*, A. and C. Black, London , 409.
3 Pickering, E.C., 1890, *American Journal of Science* (3rd series), **39**, 46.
4 Vogel, H., 1889, *Pub. Potsdam Obs.*, **7**, 111.
5 Frost, E., 1908, *Astr. Nach.*, **177**, 171.
6 Ludendorff, H., 1908, *Astr. Nach.*, **177**, 7.
7 Gutmann, F., 1968, *Publ. DAO*, **12**, 361.
8 Pease, F.G., 1927, *Proc. Astron. Soc. Pacific*, **39**, 3.
9 Curtis, H.D., 1906, *Lick Obs. Bull.*, **4**, 55.
10 Adams, W.S. and Joy, A.H., 1920, *Proc. Astron. Soc. Pacific*, **32**, 158.
11 Qian, S. et al., 2002, *Astron. J.*, **124**, 1060.
12 Savary, F., 1827, *Connaissance des temps pour l'an 1830*, 56.
13 Norlund, N., 1905, *Astr. Nach.*, **170**, 117
14 van den Bos, W.H., 1928, *Danske Vidensk. Selsk. Skrifter*, ser 8, **12**, 293.

15 Berman, 1932, *Lick Obs. Bull.*, **16**, 24.

16 Griffin, R.F., 1998, *Observatory*, **118**, 273.

17 Mason, B.D. et al., 1995, *Astron. J.*, **109**, 332.

18 Bonneau, D., 2000, *Observations y Travaux*, **52**, 8.

19 See, T.J.J., 1896, *Researches on the Evolution of the Stellar Systems*, Nichols, Massachusetts.

20 Pavel, F., 1921, *Astr. Nach.*, **212**, 347.

21 Batten, A.H. and Fletcher, J.M., 1991, *Proc. Astron. Soc. Pacific*, **103**, 546.

22 Reuyl, D. and Holmberg, E., 1943, *Astron. J.*, **97**, 41.

23 Worth, M.D. and Heintz, W.D., 1974, *Astrophys.. J.*, **193**, 647.

24 Heintz W.D., 1988, *JRASC*, **82**, 140.

25 Griffin, R.F., 2000, *Observatory*, **120**, 1.

26 McCarthy, D.W., 1983, *IAU Colloquium* 76, 97.

27 Hutchings, J.B., Griffin, R.F. and Ménard, F. 2000, *PASP*, **112**, 833.

28 Richichi, A., 2000, *Astron. Astrophys.*, **364**, 225

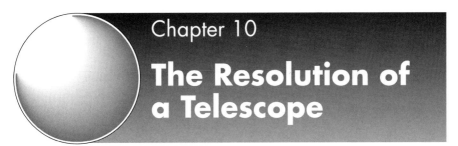

Chapter 10

The Resolution of a Telescope

Bob Argyle

The Airy Disk

The structure of an image formed by a circular aperture was first formulated by George Airy.[1] In a refractor, the effect of diffraction on the image of a star in the focal plane is to produce a series of faint concentric rings around the central disk, called the Airy disk.

The diameter of the central peak of the Airy disk is:

$$D_{Airy} = \frac{2.44\lambda f}{d}. \qquad (10.1)$$

Looking at a star, most of the light (84%) goes into this central disk inside the first dark ring. The intensity of the first bright ring is 7% of the total light contained within the star image. The second bright ring is only 3% of the total light, with the remaining 6% being distributed in the outer rings.

The Rayleigh Criterion

The theoretical diffraction image, or Airy pattern, of a star, seen in the focal plane of a perfect refracting telescope of aperture D cm, is given by the pattern in Figure 10.1. If a second star, equally bright, and close to the first is also present then two Airy disks and sets of rings are visible.

The Rayleigh criterion is defined as the separation at which the peak of one Airy disk corresponds exactly to

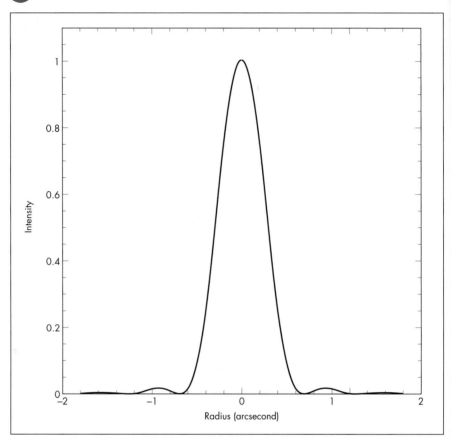

the centre of the first dark ring of the other profile. At this point the intensity in the dip between the two profile peaks drops to 73% of the intensity of either peak. In terms of the angular separation of the two stars this is given by 1.22 λ/D in radians. In terms of the resolving angular separation, θ_{res}, of the two stars

Figure 10.1. The diffraction limited image of a star in a perfect refractor.

$$\tan \theta_{res} = \frac{0.5\, D_{Airy}}{f} = \frac{1.22\, \lambda}{D}$$

Because θ_{res} is a small angle, $\tan \theta_{res} \approx \theta_{res}$, so

$$\theta_{res} = \frac{1.22\, \lambda}{D}$$

but remember that θ_{res} is measured in radians. To convert to seconds of arc, multiply by 206,265.

The power of an objective to separate double stars therefore nominally depends on both the wavelength and the diameter of the objective. For the normal eye the wavelength is that of the peak response, which is usually at 550 nm. So replacing λ in the last expression and converting from radians to arcseconds gives the Rayleigh criterion of $13.8/D$ where D is in cm.

Thus, for a 10-cm refractor, the Rayleigh criterion is $1.38''$. This corresponds to a drop in intensity of 27% in the centre of the combined profile, between the two maxima. However, it is possible to see double stars still resolved even if they are closer than this limit. This was first demonstrated, for small telescopes at least, by the Reverend William Rutter Dawes (1799–1868). Dawes says: "I examined with a great variety of apertures a vast number of double stars, whose distances seemed to be well determined, and not liable to rapid change, in order to ascertain the separating power of these apertures, as expressed in inches of aperture and seconds of distance. I thus determined as a constant, that a one-inch aperture would just separate a double star consisting of two stars of the sixth magnitude, if their central distance was $4''.56$; the atmospheric circumstances being moderately favourable."[2]

Aitken[3] points out that it is generally accepted that resolving power rests partly upon a theoretical and partly on an empirical basis. This can be seen in Figures 10.2 and 10.3. In the first, the Rayleigh criterion for a 20-cm refractor is shown with the intensity between the two peaks dropping to 73% of the maximum when the peak of one profile is $0.69''$ from the centre of the second profile. Figure 10.3 shows the situation with the Dawes limit demonstrated (the stars are $0.58''$ apart in this case). The dip between the peaks is only 3% in this case. The resolution of a double star can therefore depend on the brightness of the stars as it is easier to see a small dip in a bright image than in a faint image.

The Dawes Limit

Dawes arrived at this relationship in 1867 after tests with a large number of apertures over a number of years. Unfortunately, Dawes only had the experience of refracting telescopes and was not able to comment on the application of this relationship to reflectors, let alone modern catadioptric telescopes! In the next chapter, Christopher Taylor will argue that the Dawes

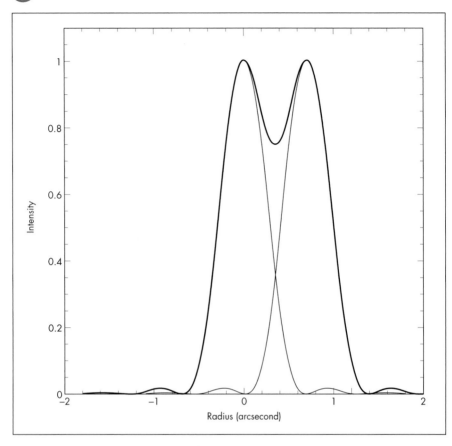

limit applies equally to reflectors at least to apertures of 30 cm.

Although the Dawes limit is an empirical limit which happens to work well for small apertures (below about 30 cm) it was clear at the turn of the last

Figure 10.2. Image profiles at the Rayleigh limit.

Table 10.1. Dawes and Rayleigh limits for various apertures

Aperture (inches)	Aperture (cm)	Dawes Limit	Rayleigh Limit
1	2.5	4″.56	5″.43
3	7.6	1″.52	1″.82
6	15	0″.76	0″.92
8	20	0″.57	0″.69
10	25	0″.46	0″.55
12	30	0″.38	0″.46
16	40	0″.29	0″.35
24	60	0″.19	0″.23
36	91	0″.13	0″.15

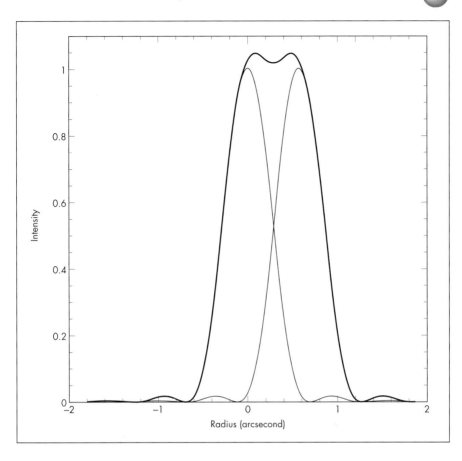

Figure 10.3. Image profiles at the Dawes limit.

century, when Aitken and Hussey were using the large American refractors, that it was not a universal limit. In particular it does not apply to unequal pairs and few attempts have been made to try and predict the performance of a given aperture in such cases. In 1914, Thomas Lewis[4] produced a number of other relationships between aperture and separating power which, he said, were more relevant to cases where the stars were either unequally bright or both faint.

Christopher Lord has recently attempted[5] to produce an empirical law which will predict the resolution of any telescope with any given aperture, central obstruction and seeing, in the case of pairs of any given magnitudes. This has been derived from observations of many binaries with a range of telescopes. For ease of use, a nomogram has been produced.[6]

The Effect of Magnification

The term "resolving power" is rather misleading as it implies that the amount of resolution depends on the magnification, which it does not. A more accurate term might be the limit of resolution or the angle of resolution. If the two images appear separate in the eyepiece then an increase of magnification should separate the images still further, assuming that the atmosphere will allow higher magnification.

The resolution of the human eye depends on the diameter of the pupil, which can vary from 1.5 mm to 8 mm depending on the individual and conditions of illumination. For double stars it is generally accepted[7, 8] that the limiting resolution is about 2–2.5′, lower than might be expected from the pupil diameter, but when the eye is fully dark-adapted, the image definition is impaired by inherent aberrations in the eye.

In terms of measuring close pairs, Couteau[9] defines a resolving magnification which makes the radius of the first dark ring equal to the visual limit for the average eye. This magnification is numerically equal to the diameter of the objective in mm, i.e. $m_r = 200$ for a 20-cm telescope. Couteau considers that the minimum useful magnification for double stars is $2m_r$, or ×400 for a 20-cm telescope.

The Effect of Central Obstructions

When a reflector or a Schmidt–Cassegrain is considered, the resolution is slightly changed by the presence of the secondary mirror. The result is to slightly reduce the size of the Airy disk and reduce the radii of the bright rings, at the same time slightly broadening the width and increasing the intensity of the rings. The result is that, for equal pairs, the reflector is as effective as the refractor until the central obstruction is greater than about 33%; but for unequal pairs the wider diffraction ring makes it more difficult to see faint stars close to bright ones. Christopher Taylor will go into this in more detail in the next chapter, which will deal with the effect of alignment and aberrations on resolution for Newtonian reflectors.

Using Aperture Masks

As we have seen above, the circular form of the telescopic image is due to the shape of the diffracting aperture. The effect of the secondary mirror of a reflector modifies the size of the Airy disk and the radius and intensity of the diffraction rings.

The use of an aperture mask has been applied in several ways to modify the imaging of a telescope to deal with particular problems in imaging double stars, in particular with binary stars such as Sirius where the companion star is very much fainter than the primary, 10,000 times as faint in fact. Unless Sirius B (also called The Pup) is near its widest separation (about 11″) it is impossible to see visually with a small telescope. This is because the glare from Sirius A spreads out to envelop the companion star.

One means of reducing the glare is to use a hexagonal aperture mask, a fact that seems to have been discovered by Sir John Herschel. The effect is to produce a six-pointed diffraction pattern, with most of the light being directed into these spikes, and the sky between the spikes, relatively near the brilliant primary star, being much darker than without the mask. E.E. Barnard[10] used this method to measure Sirius B. By rotating the mask around the optical axis, it can be used to glimpse faint companions at any position angle to bright stars.

Another form of aperture mask is the coarse diffraction grating. Used by professional astronomers to reduce the brightness of the components of double stars whilst maintaining the resolution, the grating can also be used as a basis of a simple micrometer, the principle and operation of which can be found in Chapter 14.

Experiments have been made with other shapes of aperture masks. G.B. van Albada[11] describes the use of an objective mask made from several lentil-shaped slits which were used in double-star photography on the 23.5-inch refractor at Lembang in Java. It was possible to just record the companion of Procyon (a considerably more difficult star than Sirius B) using this method.

A new application of this principle is being considered for imaging extrasolar planets close to bright stars. Whilst a sharp aperture produces a fuzzy image, it turns out that the converse is also true. By using a square aperture with a fuzzy edge, thus directing most of the light into four diffraction spikes at right angles to each other, NASA astronomers hope to find planets by direct imaging. The process of producing a fuzzy aperture is analogous to apodising where, by coating a lens with a film which is progressively thicker towards the outer edge, the effect on the Airy disk is to increase it in size but the diffraction rings are suppressed. A fuzzy-square mask should make it possible for telescopes to see Earth-like planets about five times closer to their star than with an ordinary telescope.

Below the Rayleigh Limit

Airy's definition does not mean that closer pairs than this cannot be seen. In fact, elongations of the image can be followed down to a fraction of the resolving power. Simonow[12] has tabulated the relationship between the shape of the image and the angular separation as the latter drops further below the nominal resolving power. For the 23.5-inch refractor at Lembang (Rayleigh criterion, $R = 0\rlap{.}''23$) he came up with the following:

Just separated:	$0\rlap{.}''23 = 1.00R$
Notched:	$0\rlap{.}''21 = 0.95R$
Strongly elongated:	$0\rlap{.}''19 = 0.86R$
Elongated:	$0\rlap{.}''17 = 0.77R$
Slightly elongated:	$0\rlap{.}''15 = 0.68R$
Elongation suspected:	$0\rlap{.}''14$ (minimum distance estimated)

Thus Simonow was able to detect duplicity for pairs whose separations were about 0.7 of the Rayleigh criterion. Simonow extended his discussion of resolving power to include other combinations of magnitude and magnitude difference.

Paul Couteau has also discussed this subject in depth and obtains slightly smaller figures than Simonow for the 50-cm refractor at Nice. He claims that the limit at which stars can be seen as double is 0.14″ or half of the Rayleigh limit for this aperture.

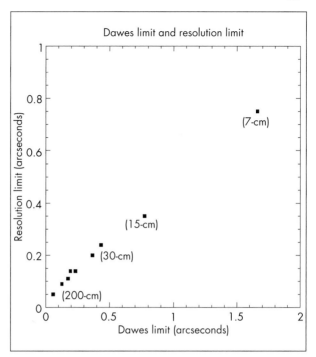

Dawes limit and resolution limit

(7-cm)

(15-cm)

(30-cm)

(200-cm)

Figure 10.4. Plotting the Rayleigh limit against the limit of duplicity.

An investigation into this by the writer has produced the graph in Figure 10.4 which shows the least angular separation at which close double stars in various apertures have appeared to be just distinguishable from a single image and it shows a surprisingly good correlation from the smallest to the largest aperture considered.

For a list of close pairs suitable for testing the resolution of a telescope see the tables in Chapter 2.

Small Apertures

Jerry Spevak, observing from Canada, has recently carried out an investigation into the resolution limit of a small telescope using double stars from the *Hipparcos* and *Tycho* catalogues. He worked through the pairs without advance knowledge of Δm, and exact separation with the catalogue being checked only after noting the appearance of close pairs.

He found that the images of double stars can be divided into four categories, depending on separation: separate, touching, notched and elongated. These

Table 10.2.

Pair	Magnitudes	Separation	Images
STF 65	8.0, 8.0	3″.1	separate
STF 1905	9.1, 9.2	3″.0	separate
STF 1284	8.2, 9.7	2″.5	separate
STF 2845	8.1, 8.3	2″.0	separate
STF 2807	8.7, 8.8	1″.9	touching
STF 2509	7.5, 8.3	1″.7	notched
STF 2843	7.1, 7.4	1″.5	notched
STF 3062	6.5, 7.4	1″.5	notched
STF 3017	7.7, 8.6	1″.4	notched
BU 1154	8.6, 8.8	1″.2	notched
STT 50	8.5, 8.6	1″.1	barely notched
STF 2054	6.2, 7.2	1″.0	barely notched
STF 2438	7.1, 7.4	0″.8	elongated
STF 2	6.8, 6.9	0″.7	elongated

classifications are fairly self-explanatory. Examples of close pairs are given in Table 10.2 along with the observed appearance and relevant data from *Hipparcos*.

The telescope for this project is small but of high quality (Figure 10.5). The small aperture has helped reduce atmospheric effects. It is a 70-mm f/6.8 apochromat on a very sturdy mount and, using powers of 137 and 200, each pair is examined for at least a minute. Even the closest pairs tend to "jump" out in a few seconds but the extra time is useful for detecting doubles whose components have a large difference in brightness.

Seeing

An Airy disk surrounded by several stationary diffraction rings is, alas, a rare telescopic sight – the presence of the Earth's atmosphere sees to that. In addition to absorbing the incident starlight, it also causes the star images to change in size (seeing), move about (wander) and to change in brightness (scintillation). Another significant effect which is better seen in larger telescopes at high magnification is the appearance of speckles, which are diffraction-limited images of the Airy disk and explained in more detail below.

Essentially, in a small telescope, aperture limits the resolution. With a large aperture the seeing limits the

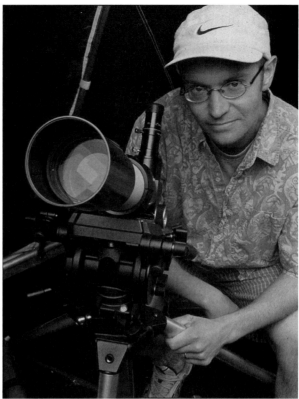

Figure 10.5. Jerry Spevak and his 70-mm apochromatic refractor.

resolution. Many observers try to quantify conditions of atmospheric steadiness and clarity by reference to a numerical scale. There are several scales of seeing and whether the numerical value of seeing increases or decreases as seeing gets better is purely a matter for personal choice. Aitken and van den Bos, for instance, each used a scale of 1 = worst to 5 = best with the occasional use of a + sign to indicate "slightly better than" as in 2+. It is difficult to justify a scale that goes from 1 to 10, for instance, because it would be difficult to be that specific about what is, after all, a very subjective parameter.

The performance of a telescope on double stars can be improved by considering some of the following points:

(a) Don't take a telescope out of a warm house into a cold garden and expect to see point-like images straightaway. The telescope must be given time to reach the temperature of the night air. This goes for the eyepieces as well.

(b) Don't be put off by a little mist or haze or even thin cloud. The atmosphere on these occasions is usually calm and can result in good seeing.

(c) If housed in an observatory, open the dome as soon as is practicable. Just after sunset is not too soon. Keep the dome closed during the day but allow a little air circulation if possible.

(d) Don't observe from surfaces which absorb a lot of heat. Grass is more preferable to concrete.

(e) Don't use a magnification which is clearly too high for the state of the atmosphere. If the images do not show disks, wait until things have improved. If the star you are after cannot be resolved, switch to a backup programme of wider pairs but always be prepared to take advantage of good seeing when it occurs.

(f) Plan your observing so that your target stars are as close to the zenith as possible when you observe them.

(g) Make sure that the telescope optics are as well adjusted as possible. For reflector users see Chapter 11 for advice on how to improve optical performance.

References

1 Airy, G.B., 1835, *Cambridge Phil. Trans.*, **5**, 23.
2 Dawes, W., 1867, *QJRAS*, **35**, 154.
3 Aitken, R.G., 1962, *The Binary Stars*, Dover.
4 Lewis, T., 1914, *Observatory*, **35**, 372.
5 Lord, C.J.R., http://www.brayebrookobservatory.org
6 Lord, C.J.R. (see Haas, S., 2000, *Sky & Telescope*, **102**, 118).
7 Sidgwick, J.B. (rev. Muirden), 1979, *Amateur Astronomer's Handbook*, Pelham.
8 Nicklas, H., 1994, *Compendium of Practical Astronomy* (ed. G.D. Roth), Springer.
9 Couteau, P., 1981, *Observing Visual Double Stars*, MIT Press.
10 Barnard, E.E., 1909, *Astr. Nachr.*, **182**, 4345.
11 van Albada, G.B., 1958, *Contributions of Bosscha Observatory*, 7.
12 Simonow, G.V., 1940, *Annals of Bosscha Observatory*, **IX**, Pt 1.

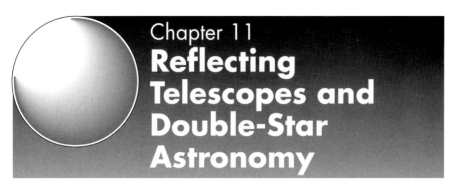

Chapter 11
Reflecting Telescopes and Double-Star Astronomy

Christopher Taylor

Reflectors versus Refractors, Optical Principles

Even a cursory reading of the literature of visual double-star astronomy is sufficient to show that the field has long been heavily dominated by the refractor, which remains the instrument of choice for many visual observers. It is not, indeed, hard to find statements backed by the highest authority alleging that for this type of observation a reflector must be of substantially larger aperture to match the performance of a refractor of given size. For instance, van den Bos stated that a reflector must have a linear aperture 50% greater than that of an equivalent refractor. There is, however, no basis whatever in optical theory for such claims, nor, as will shortly be seen, do actual results at the eyepiece sustain this perception of the reflecting telescope as second-class citizen. This chapter will demonstrate that, and how, a reflector of good optical quality, maintained in proper adjustment, can be fully the equal aperture-for-aperture of the best refractor, matching the latter's resolution to the uttermost limits of visual double-star astronomy, at least on fairly equal pairs. It is not amiss to recall at this point that the study of binary stars was founded by Herschel with reflecting telescopes and that its current limits have largely been set by recent observations with reflecting systems, both in terrestrial speckle interferometry and in the *Hipparcos* orbital observatory.

Present purposes would not be served by entering into the minutiae of the apparently interminable debate over the relative merits of the two classes of instrument, but there *are* important differences between their respective imaging properties, and handling characteristics in real observing conditions, which must be recognised by any observer who aims to push telescopic performance to its limits. There are, accordingly, a few fundamental optical principles which must be borne in mind as the essential context for what is said later in this chapter specifically about reflecting telescopes. In particular, given the myths, misconceptions and dubious anecdotal evidence common in the "refractor versus reflector" debate, it seems appropriate to begin by stating clearly what are *not* the reasons for significant differences between the two types–not, at least, so far as double stars are concerned.

One such notion holds that residual chromatic aberration is a serious limitation to the defining power of refractors with simple doublet objectives, and that the reflector therefore has a marked superiority in this sense. That there is, in fact, no theoretical justification for this view in the case of any refractor of sufficiently long focus to be used for high-resolution imaging (say f/10, at least, for smaller apertures, rising to f/18 or so for large instruments) has been known at least since the work of Conrady.[1] It was shown there that moderate levels of defocussing such as may be induced by the secondary spectrum in such a refractor, that is up to one quarter or even one half of a wavelength phase-lag, does not significantly alter the diameter of the Airy disk formed by the telescope, despite its intensity declining noticeably. Effectively, the chromatic dispersion of focus is lost within the depth of focus naturally allowed by the wave theory; this is the reason why image definition is so good in refractors despite secondary spectrum. The result is that resolution on high-contrast targets such as double stars is fully maintained, even if some low-contrast fine detail may be lost in planetary images. That this conclusion is fully borne out by practical experience is convincingly demonstrated by the magnificent achievements in high-resolution double-star astronomy of the best visual observers using the big refractors: one only need think of the Lick 36-inch regularly reaching 0.1″ in the hands of Burnham, Aitken and Hussey. Indeed, one of the greatest of recent observers of visual binaries, Paul Couteau, seems from the remarks in his well-known book[2] to consider the secondary spectrum of refractors to be a positive advantage. Clearly, three-colour or apochromatic correction, whatever its benefits for the use of relatively short-focus instruments in planetary imaging, is for the double-star observer an expensive and dispensable luxury – the classical long-focus doublet objective is more than equal to the task required.

The effects of central obstructions, often alleged to degrade imaging quality of reflectors quite seriously compared with that of refractors, can similarly be dismissed. By blocking a small central patch of the incident wavefront, the secondary mirror of a reflector removes a minor portion of the light from that process of mutual interference at focus, which otherwise produces a standard Airy diffraction pattern. The result is that an equal amount of light which would previously have interfered, constructively or destructively, with this obstructed portion in the process of image formation must now be redistributed in the Airy pattern. It follows on simple grounds of energy conservation that the amount and location of this redistribution of light in the image is essentially identical with the intensity distribution in the image which would be formed alone by just the light that has actually been blocked – a statement familiar to all students of diffraction theory as Babinet's principle (the Complementary Apertures theorem).[3] One can immediately see from this that, for the fairly small central obstructions of most reflectors, the amount of light redistributed in the image must be very small and, as the point-spread function of the obstructed central zone is very much wider than that of the full aperture (in inverse ratio to their diameters), this small amount of light is deflected from the Airy disk into the surrounding rings. It is, therefore, quite impossible for a secondary mirror blocking, say 5% of the incident light, to cause a redistribution of 20% of what remains from diffraction disk to rings, a change which would itself be near the limits of visual perception even on planetary images. This is the case of a "22.4% central obstruction" in the linear measure usually applied to discussions of this issue, and even this is decidedly on the large side for most Newtonians, at least, of f/6 and longer.

Central obstructions are not in fact the only possible cause of excess brightness in the diffraction rings nor, probably, indeed the most important single cause in the vast majority of reflecting telescopes. The effect of deviation of light from the Airy disk into the rings is quantified by the Strehl ratio, a parameter commonly used as a measure of imaging quality and as a basis of optical tolerance criteria, which is the peak central intensity of the diffraction pattern actually formed by an instrument, expressed as a fraction of that of the ideal Airy pattern appropriate to the case. The essential point here is that *any* small deformations, W, of the wavefront converging to focus, whether arising in the telescope from surface errors of the optics or from aberrations, will reduce the Strehl ratio and so cause the kind of effect commonly attributed to "central obstructions". According to Maréchal's theorem, this deviation of light from disk to rings is proportional to the statistical variance (mean square) of the wavefront deformations, W, thus:

$$\text{Strehl ratio} \cong 1 - \left(\frac{4\pi^2}{\lambda^2}\right).\text{var}(W).$$

This approximation holds for W values up to about the Rayleigh "quarter-wave" tolerance limit and in that range is independent of the nature of the wavefront deformations. More than half a century after Maréchal's discovery it is extraordinary how little-known this fundamental result[4] appears to remain in the practical world of telescope users and makers.

In particular, it turns out that spherical aberration (SA) in small doses mimics the diffraction effects of central obstructions particularly closely, putting extra light into the rings, while leaving the size of the Airy disk unaltered. With SA just at the Rayleigh limit, Maréchal's theorem shows that the Strehl ratio will already have dropped to 0.8, an effect fully as large as that of a 30% central obstruction. The conclusion is that, unless a reflector is of very high optical quality and very precisely corrected, or has an exceptionally large secondary mirror (or both), any effect of the central obstruction will be swamped by that of SA, to say nothing of other aberrations and optical errors. This is particularly significant in view of the prevalence of residual SA in reflecting telescopes: plate-glass mirrors tend to go overcorrected in typical night time falling temperatures, so older optics even from professional makers are often undercorrected, deliberately; the absence of a simple null-test for paraboloids, and the acquired skill necessary to interpret accurately the results of the Foucault test at the centre of curvature, mean that amateur-made mirrors are often only very approximately corrected; and Cassegrain systems, such as the ubiquitous SCT compacts, which focus by moving one of the main optical elements, necessarily introduce correction errors for all settings except that in which the principal focus of the primary mirror coincides exactly with the conjugate focus of the secondary. For a very interesting field survey of the effects of residual correction errors on performance of reflectors see reference 5. A further point here is that SA is proportional to (aperture)^2/focal length, so the claim that the "cleaning up" of the image in a typical reflector by use of an off-axis unobstructed aperture proves that the secondary mirror is responsible for the less-than-ideal image at full aperture is obviously a misinterpretation of the evidence: simply by stopping down, both SA and "seeing" effects are drastically reduced, naturally giving rise to the observed changes in image quality.

These conclusions are entirely vindicated by practical experience. In the 12.5-inch (0.32 m) f/7 Newtonian with whose star images this author has been intimately familiar since the 1960s, increase of the normal 16% central obstruction to 32% has no perceptible effect on the diffraction image of a first magnitude star, although the brightening of the rings has become very obvious at 60% obstruction. Again, a deliberate trial of this question was made by side-by-side star tests, on the same bright star, of a 4-inch refractor and a 6-inch Newtonian having 37% central obstruction. With both instruments showing a beautifully defined Airy pattern at ×200, the greater relative intensity of the rings in the reflector was so small as to be barely detectable even after many rapidly alternated comparisons. It should be noted that even this rather large obstruction only stops about 1/7 of the incident light.

In short, the unavoidable presence of a central obstruction in most reflectors does *not* limit their resolution, or make it inferior to that of refractors of equal aperture. On the contrary, by stopping out the centre of the mirror, the mean separation of the points on the incident wavefront is increased, thereby decreasing the size of the Airy disk which arises from their mutual interference, so the resolving power of a reflector on fairly equal double stars is actually greater than that of a refractor of the same aperture, other things being equal. In truth, this last effect is almost negligible for central obstruction much below 50% but it may surprise some readers to learn that for the highest resolution on equal pairs this author deliberately stops out the central 72% of the telescope's aperture – a 9-inch central obstruction on a 12.5-inch reflector! (None of the double-star results given later were dependent on this trick, however.) Of course, such doubles are extreme high-contrast targets and therefore react quite differently to such treatment, compared with planets or even unequal double stars, whose resolution would be seriously impaired by this tactic.

To bring this discussion to its conclusion, the real differences between refractors and reflectors which are important for high-resolution imaging of double stars are very simple and very fundamental: refractors refract, while reflectors reflect and refractors do this at four (or more) curved optical surfaces as against only one in a Newtonian. These two facts are so obvious that they are often ignored but they are, far more than any other factors, truly the crux of the matter in comparing the optical performance of the two main classes of instrument.

That image-formation is, in the one case, by refraction, and, in the other, by reflection has radical implications for the relative immunity of the refractor from image degradation due to surface errors of the optics, whether arising from inaccuracy of figuring, thermal expansion or mechanical flexure. Thinking in wave terms, one can say that the function of a telescope's optics in forming a good image of a distant star is simply to cause rays from all points of the plane wavefront incident on the aperture to travel exactly the same number of wavelengths (optical path-length) in arriving at the focus, so that they may interfere constructively there and form a bright point of light. That is all there is to image formation in the wave theory, whether by refraction or reflection (and this is precisely why results like the Airy pattern and Maréchal's theorem arise) – arrival in phase of all rays at focus. The refractor achieves the necessary phase delay of the near-axial rays, relative to the peripheral rays which must follow a longer route to focus, by intercepting them with a greater thickness of dense optical medium to equalise axial and peripheral optical path lengths. That is to say, the telescope uses a convex lens. The reflector attains exactly the same result by bouncing the axial rays back up to focus from further down the tube than the peripheral rays, that is, it uses a concave mirror.

It immediately follows that this differential phase-delay, and hence quality of image, is dependent on the *thickness* of the objective at any point relative to that at its edge, in a refractor, but on actual longitudinal *position* of the mirror surface relative to the edge, in a reflector. Further, errors of glass thickness in the first case only cause optical path-length errors $(\mu - 1)$ times, or approximately half as great, while errors of surface in the second case are doubled on the reflected wavefront, as such errors are added to both the to and fro path length. Consequently, to achieve any particular level of wavefront accuracy var (W), and thus image quality (cf. Maréchal's theorem, above) in a reflector requires optical work roughly four times more accurate than in the case of a refractor and, for exactly the same reason, the latter is about four times less sensitive, optically, to uneven thermal expansion of its objective. Lastly, because mechanical flexure does not alter thickness of an objective in first approximation, while it has an immediate and direct effect on the local position of surface elements of a mirror, refractors are hugely more resistant to the optical effects of flexure.[6,7]

That refractors share the work of focussing light between at least four curved surfaces, compared to only one in a Newtonian, is equally fundamental and takes us to something which will be the central theme of the next few pages: optical aberrations and their avoidance or management. The requirement that a curved mirror surface return all rays incident parallel to the optical axis to focus with equal optical path lengths, so forming a fully corrected image there as discussed above, is alone sufficient to determine uniquely the form of that surface. A very simple geometrical construction shows that the mirror must be a paraboloid of revolution. In other words, the requirement that axial aberrations, specifically SA, be zero defines the optical configuration uniquely and leaves no adjustable parameters free for reducing or eliminating off-axis aberrations (apart, trivially, from the focal length). The result is that all reflectors, Newtonian, Herschelian, or prime focus, having only one curved optical surface, necessarily suffer from both coma and astigmatism. Unless other adjustable optical surfaces are introduced into the system, nothing can be done to mitigate the full force of these off-axis aberrations and, as will be seen in the next section, coma severely limits the usable field of view of all paraboloid reflectors and makes them hypersensitive to misalignment of the optics (collimation errors). A refractor objective, by contrast, possesses at least four independently adjustable curvatures and opticians have known since the time of Fraunhofer how to use this freedom to eliminate both the axial aberrations and coma, in the so-called aplanatic objective. (The need for multiple-surface adjustability to minimise aberrations is, of course, the reason why all short-focus wide-field imaging units such as camera lenses and wide-field eyepieces must have four or more components.) Most quality refractor objectives are nearly or quite aplanatic, leaving only astigmatism as the factor limiting field of view, a very much less serious constraint which leaves most refractors with a far larger field of critical definition and far less sensitivity to collimation errors than all Newtonians, at least. Compound reflectors such as Cassegrains or catadioptrics represent a halfway stage in this sense between Newtonians and aplanatic refractors but most of these pay the price of decreased (rarely eliminated) coma in increased trouble from SA. Coma arising from miscollimation in reflectors is perhaps the most obnoxious of all aberrations to the double-star observer, as it rapidly destroys the symmetry and definition of the star image: even a quarter wave of coma, that is just at the Rayleigh tolerance, is quite sufficient to make the diffraction rings contract into short, bright arcs all on one side, an image distortion quite unacceptable for critical double-star observation – see Figure 11.1.[8]

What all of this amounts to in practice is that a reasonably well-made Fraunhofer achromat is a hugely more robust instrument than a typical reflector in the face of the thermal variations, mechanical flexure and shifting collimation which commonly arise in real observing conditions, and so can be relied upon far more than the comparatively delicate, fickle reflector to deliver critical definition at a moment's notice with minimal cosseting and adjustment. It is also more likely to meet the optical tolerances necessary for such diffraction-limited performance. *These* are the reasons why the refractor has so often been the first choice for observers of close visual binaries.

However, as will be seen shortly, none of this implies an inevitable inferiority of the reflector in this field of

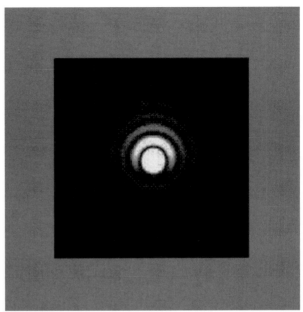

Figure 11.1. A quarter wave of coma.

astronomy, for good optics and proper management of the instrument will easily hold in check all those adverse factors to which the reflector is more sensitive, to an extent quite sufficient to deliver star images equal to any seen in a refractor. (With the possible exception of some enhancement of the diffraction rings in reflectors exhibiting residual SA. If this is the only fault, the telescope will perform just as well on equal double stars but faint companions may be swamped. For this reason, a good refractor will often outperform a reflector on contrasted pairs even when the two instruments are absolutely matched on equal doubles.) All the supposed optical defects of the reflector are removable or fictitious and, of course, a good 0.3-m reflecting telescope is a far less expensive item than an equally good 0.3-m refractor! For reaching the observational limits, however, the unrelenting emphasis must be on quality optics and their proper management, in particular to maintain accurate collimation so that all high-power images may be examined truly on axis, free of the dreaded coma. This is the subject of the next few sections. What follows is largely based on experience with a Newtonian reflector, with which this author has done most of his double-star astronomy, but results comparable with those reported here are probably within reach of good longish-focus reflectors of virtually any type, given the same aperture.

Coma and Astigmatism

For a paraboloid mirror, the angular expansion of an image due to coma, ξ, depends on the telescope diameter D and focal length f by the following relation:

$$\xi = \frac{3}{4}\left(\frac{D}{2f}\right)^2 \tan \theta \quad \text{(in radians)}$$

where θ is the angle of the incident ray to the optical axis. The angular expansion of the image due to astigmatism is:

$$\sigma = \left(\frac{D}{2f}\right) \tan^2 \theta.$$

For the case near the optical axis $\tan \theta \simeq \theta$ in radians and so in this case these relationships simplify to:

$$\xi = \frac{3\theta}{16F^2} \quad \text{and} \quad \sigma = \frac{\theta^2}{2F} \tag{11.1}$$

where $F = f/D$ is the focal ratio.

It is more convenient in practical terms to express the angular distance off-axis in arcminutes and the aberrations in arcseconds, when the first result becomes

$$\xi = \frac{11.25\ \theta'}{F^2} \tag{11.2}$$

This is in close agreement with Bell.[9] Since ξ is linear in θ, whilst σ is quadratic, it follows that, on moving off-axis, coma is always the first aberration to appear and that, for the small θ values with which we are concerned, astigmatism is generally negligible compared with coma for all except very extreme focal ratios. Their ratio is $\sigma/\xi = 8F\theta/3$ which, for example, only reaches unity at f/6 rather more than 3.5° off-axis and is ≤ 0.1 at this f-ratio out to $\theta = 21.5'$. A Newtonian showing astigmatic star images is, therefore, either grossly misaligned – to the point that the reflection of the diagonal in the main mirror will be wildly eccentric – or has a badly distorted optical figure.

Impairment of Resolution/Image Quality

Bell[9] says that resolution will be noticeably impaired if the off-axis aberrations (which the image may exhibit even at the centre of the field due to imperfect collimation) are approximately equal to the empirical resolution limit $4''.56/D$ (Dawes limit). Despite some statements to the contrary in the literature there is no doubt whatever that this criterion is true, as is fully borne out in my experience by a good deal of very exacting double-star observation at the 0.3–0.4″ level with an f/7 mirror of 12.5 inches diameter. Thus to achieve full resolution we must operate on or near the true optical axis, at $\theta \le \theta_{max}$ where θ_{max} is the angular displacement off-axis at which $\xi + \sigma =$ the Dawes limit. In view of the comments above regarding the smallness of σ, we can approximate this condition closely by the simpler $\xi =$ Dawes limit (first-order approximation, valid for all normal f ratios) which, with all angles in radians, is

$$\frac{3\theta}{16F^2} = \frac{2.21 \times 10^{-5}}{D}$$

where D is the aperture in inches. Hence

$$\theta_{max} = \frac{1.18 \times 10^{-4}\ F^2}{D}$$

or in arcminutes:

$$\theta_{max} = \frac{0.405\ F^2}{D} \tag{11.3}$$

This angle is the limitation to the field of critical definition centred on the optical axis and is, therefore, also a measure of the maximum angular error which can be tolerated in collimation of the telescope's optics, specifically, in the squaring-on of the main mirror. The noteworthy point here is the extremely small value of this angle even for unfashionably long Newtonians (which, of course, are far better in this sense since θ_{max}

$\propto F^2$), far smaller in fact than the attainable tolerance of the methods of collimation in general use: for the 12.5-inch at f/7.04, the formula gives $\theta_{max} = 1.6'$ – a value again fully borne out by my observational experience. (This implies a maximum field of critical definition of 3.2', compared with an actual field of 2.4' on this instrument at the power used for subarcsecond pairs (×825).) In fact, I would say that for really critical double-star work right at the limit of resolution on a Class I or II (Antoniadi) night, aberrations become quite noticeable even at half this level, so reducing θ_{max} to 0.8', i.e. 48″ – about the size of Jupiter's disk! Furthermore, as this angle varies as the square of the f-ratio, the modern generation of short-focus Newtonians are at a huge disadvantage here and it is probably true that no Newtonian at f/5 or below will ever, in real observing conditions, reach anything approaching its limiting resolution. Even if one can guarantee the hyperfine collimation tolerance demanded (and in my experience these instruments are used most of the time with squaring-on checked only to ± 0.5° or worse, i.e. only the first approximation to collimation is carried out), the objects observed will almost never lie in this minute axial patch of the field of view.

Under what conditions will the first-order approximation above for θ_{max} be valid? We may reasonably say that astigmatism is negligible if, say, $\sigma/\xi \leq 0.1$ and this imposes the condition that $F\,\theta_{max} \leq 3/80$, which on substituting the first-order approximation for θ_{max} (in radians) yields $1.18 \times 10^{-4}\,F^3/D \leq 3/80$. Thus the mathematics is self-consistent, and the first-order result for θ_{max} is valid, if and only if $F^3/D \leq 318$. For the $12\frac{1}{2}$-inch telescope this parameter has the value $F^3/D = 27.9$ – well within the "coma-dominated" regime. In fact, there is no focal ratio of Newtonian likely to be encountered in ordinary astronomical use, in which the off-axis limitation to the field of critical definition is due to anything other than the onset of essentially pure coma.

It is worth bearing in mind a few numerical values of this field, $2\theta_{max}$, as given by equation (11.3) for some common Newtonian configurations: 8.6' for a 6-inch at f/8; 3.6' for an 8-inch at f/6; 2.0' for a 10-inch at f/5. Equation (11.2) then implies that at the edge of a field *n* times wider than this, the aberration will be *n* times larger than the Dawes limit.

Practical and Observational Consequences

Of prime concern here is not the issue of obtaining the largest possible field of view from the telescope at full resolution, since wide-field observation is, almost by definition, not high-resolution imaging. In any case, most of us have to make do with the fixed F and D of the telescope we have and are, therefore, stuck with the fixed θ_{max} value those imply. The real issue for practical observing, if the telescope is to be used as a serious optical instrument and not merely as a crude "light bucket", is that of *sufficiently accurate collimation of the optics* to guarantee maximum image quality and full, unimpaired resolution somewhere (preferably the centre!) in the field of an eyepiece of sufficient power to reveal that resolution to the eye. If, through failure of collimation, the optical axis of the primary mirror falls outside such a field by more than θ_{max}, the telescope will never reach its limiting resolution however good the "seeing" may be and even this is a hopelessly sloppy criterion since it allows nothing for the aberrations of the eyepiece when used far off-axis. The matter is certainly not trivial as typical fields of these very high-power eyepieces are only of the same order of magnitude as θ_{max} itself.

The usual collimation procedure[10] of looking into the telescope in daylight through an axial pinhole and centring/rendering concentric the reflections of the main mirror in the diagonal and of the diagonal in the main mirror will, if carried through carefully, bring the optical axis into coincidence with the centre of the eyepiece field to within a tolerance of order 10′. At this point, the telescope, if of good quality, will very likely yield quite pretty and satisfying images even of planets at moderately high powers (approximately 20 per inch of aperture) and stars will appear round or point-like up to about these magnifications; it will not, however, reach the limiting resolution for that aperture, falling short of this by a factor of 2 or more in all probability. This is well illustrated by a typical experience with the author's 12.5-inch. After full collimation on 1996.80, the telescope completely split and separated γ^2 Andromedae (OΣ 38) at ×825 in good but not perfect seeing, when the pair was at 0.50″. A few nights later, after a hurried setting-up in which it had not been possible to complete the final stages of collimation, there was no trace of the companion visible at that power in the same instrument, despite superlative seeing and the star on the meridian. The residual aberrations which blotted out the little star on this occasion were nevertheless still so small as to be completely inappreciable in planetary images; Saturn that night was magnificent at ×352.

To go beyond this sort of 30–50% performance there are two further stages which must be completed, what one might call "fine collimation" and "hyperfine collimation", the first a refinement of the usual daylight procedure, the second using night-time star tests. No progress can be made on either of these unless the telescope is fitted with fine adjustment screws controlling the squaring-on of the main mirror cell, which are themselves driven by controls within comfortable reach of the eyepiece. Note that it is vital that the observer is able to alter the attitude of the main mirror at will *while* looking through the eyepiece. Given how very simple it is to contrive this on the majority of Newtonians, it is remarkable how few instruments, commercial or home-made, are fitted with the necessary gear. Having equipped the telescope with this, one can proceed with daylight fine collimation. Mark the centre of the main mirror surface (the pole of the paraboloid) with a round spot at least 1/8 inch across – Tippex is very suitable – the precise size is of no importance but what is absolutely vital is that it be plainly visible from the eyepiece drawtube, be exactly concentric with the pole of the mirror and be fairly accurately circular. Point the telescope at the daylit sky and look along the axis of the drawtube, accurately defined by a "dummy" eyepiece or high-power eyepiece from which the lenses have been removed. Having made the usual adjustments to the diagonal, use the mirror-tilt fine adjust screws to move the reflection of the diagonal in the main mirror until its centre falls exactly on that of the Tippex pole-mark. This should be done by winding the adjust screws, and hence the reflection of the diagonal, to and fro repeatedly while watching through the drawtube, until absolutely satisfied of complete concentricity of diagonal-reflection and pole-mark, so far as the eye can judge. This will probably have taken collimation to within 2′ or 3′ of target. All of this assumes the mounting of the diagonal to be rigid, without perceptible play; the small shifts in position (e.g. rotation about the optical axis) of a floppy diagonal can easily introduce randomly changing collimation errors of 10′ or more, so defeating all one's best efforts. Nor can it be assumed that collimation is an infrequent necessity, let alone a once-for-all ritual; even a permanently mounted instrument is subject to frequent shifts and distortions (mechanical flexure, thermal expansion and contraction, etc.) at the arcminute level and my personal experience is that serious attempts on subarcsecond double stars require recollimation at each observing session. However, once in the habit of it, the process takes only a couple of minutes – hardly a major chore.

For the final, hyperfine stage one has to wait for a class I or II (Antoniadi) night, to push the telescope to its absolute limits. This stage is, of course, only relevant to observing on such nights, in any case. Charge the telescope with a power of ×50 to ×80 per inch of aperture (e.g. $\frac{1}{4}$-inch eyepiece and Barlow pushed well in) and focus on a second or third magnitude star. An immediate test of the quality of the telescope is that even at this power the star should come crisply to focus so that the central disk is almost pinprick-like (this may well be surrounded by a fainter and much larger fuzzle of instrumental and atmospheric origin but ignore that to start with) and unless the instrument is of uncommonly long f-ratio there will be virtually no depth of focus – the tiniest displacement of the eyepiece in or out will noticeably de-focus the star image. (The theoretical depth of focus is $\pm 8F^2 \Delta\lambda$ where $\Delta\lambda$ is the maximum tolerable wavefront deformation arising from malfocus.[1] If we adopt the Rayleigh tolerance limit $\Delta\lambda = \lambda/4$, this becomes $\pm 2F^2\lambda$: e.g. $\pm 99\lambda$ at f/7, which is just over 0.05 mm.)

It is, however, the diffraction rings which are far the most sensitive indicators of image degradation due to atmosphere, bad optics or imperfect collimation, which is why one so rarely sees the ideal Airy pattern of the books under real field conditions – and which, rather than the central disk, are therefore used for monitoring hyperfine collimation. The rings are, in particular, extremely sensitive to coma due to miscollimation and will show a very pronounced lopsidedness at a far lower level of maladjustment than is needed to make the central disk go visibly out of round. The result in a Newtonian can be a really quite serious loss of resolution as all the light previously distributed evenly and

symmetrically around the rings is dumped into a collection of much brighter short arcs all to one side, creating a sort of false image several times the size of the
Airy disk. It seems that this degree of comatic distortion occurs at about the 2–3′ level of collimation error
one can hope to achieve at the fine collimation stage –
depending, of course, on f-ratio but that is my experience at f/7.

Assuming that fine collimation has been carried out
with sufficient care and that the optics are of reasonable quality, a close look at the halo or fuzzle surrounding the main star image should reveal that it is
at least partly composed of very roughly concentric
bright arcs vaguely centred on the star disk. In a
Newtonian of typical proportions there are likely to
be three or four quite bright arcs (often a lot brighter
than the theoretical Airy ring pattern, as noted in the
first section of the chapter) and you will be doing
extremely well at this stage to see them as arcs of
more than about 120°. Unless the night is a true class I
(i.e. very rarely at most sites) the rings are not easily
seen on full aperture the first time one tries this; they
will be fragmented, distorted crinkly-wise and constantly on the jitter. If previous adjustments have
brought the telescope within 2 or 3′ of true, you will
be operating by now well within the coma-dominated
regime discussed earlier and an idealised version of
what you will see (ignoring atmospheric interference)
is shown in Figure 11.2. What you almost certainly will
not see is a complete set of circular rings.

Coma in a Newtonian off-axis is external; that is to
say the light of the diffraction rings is displaced to the
side furthest away from the optical axis. The remedy to
the state of affairs shown here – the final hyperfine collimation – is therefore simple (in principle!): while
keeping close watch through the eyepiece, wind the
fine-adjust controls on the main mirror very slowly
so as to displace the distorted image (Figure 11.3),
re-centring the star in the field as this adjustment proceeds. It may well be that the outer arcs will disappear
during this process but the important thing is that the
innermost arc should expand tangentially so as to
encircle the central image as a complete ring of
uniform brightness. If that state is achieved, you will be
in the fortunate position of having a telescope which
will reveal detail right down to its diffraction limit –
atmosphere permitting!

Figure 11.2. The effect of slight miscollimation in a reflector.

It should now be evident why such insistent emphasis was placed earlier on the need for the collimation controls to be within comfortable reach of an observer actually looking through the instrument, for without

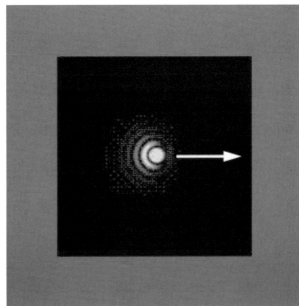

Figure 11.3. Correcting the miscollimation.

such provision fine collimation will obviously be almost impossible and, in view of the very high powers needed during this stage, hyperfine collimation will be absolutely out of the question. This last stage of collimation, using the structure of star images, must be conducted with the telescope at full aperture but it may take some initial practice for less experienced observers to see the relevant details of the diffraction pattern. Readers unaccustomed to such high-power observation and to the appearance of the Airy rings may find it helpful to follow the suggestions made in the section below on "How to See the Diffraction Limit of any Telescope" before attempting star tests and hyperfine adjustment.

On the Possible Occurrence of Astigmatism in Star Tests

The plain fact is that there shouldn't be any. Provided that the optics themselves are of true figure, coma is the only image defect which can occur for small deviations off-axis due to imperfect collimation of a Newtonian. By the time that even rough collimation has been done, the instrument should be well within the coma-dominated regime, as explained above. Conversely, to make astigmatism dominate, the telescope would have to be miscollimated by an angle of order $\theta = 3/8F$ which is huge compared with the alignment tolerances discussed above. At this point the image distortions due to off-centring would be huge themselves – stars would appear all sorts of curious shapes even on the lowest powers and resolution would be degraded to tens of arcseconds – and the crudest of rough collimation by eye would eliminate the problem. In other words, *small image distortion in a Newtonian due to small errors of collimation is never astigmatism.*

If, nevertheless, the star image during hyperfine collimation looks fixedly like this (in order of increasing badness):

Figure 11.4.

 or this : or this : or this :

then you *have* got a small dose of astigmatism. As it can't be due to miscollimation, it must be due to distortion of figure in the optics but remember that there are four components to the optical train: main mirror, diagonal mirror, eyepiece and your eye. It should be quite easy to determine which of these is responsible for the problem, since all except the diagonal can be rotated about the optical axis without affecting the collimation: whichever rotating component carries the axis of symmetry of the cruciform image with it, is the villain of the piece and has a distorted astigmatic figure. If this does turn out to be the main spec, it is still not cause for despair since the condition may be temporary and remediable and, in any case, if it is only as bad as the first diagram above it will have negligible effect on telescopic resolution and one can comfortably live with it, even if permanent, i.e. the telescope is still a good one. It should be noted that the machine-generated images in Figure 11.4 are something of a theoretical ideal, as they have been computed only for exact paraxial focus. In reality, astigmatism is more likely to be noticed as a distinct elongation of the star disk when slightly out of focus, this elongation reversing on passing through the focal point. This is the most characteristic symptom of astigmatism and is very pronounced even in the first case depicted above, in which the focal star disk remains virtually unaffected.

Temporary astigmatic distortion of the main mirror can be due to a variety of causes but principally three: uneven thermal expansion/contraction in changing temperatures, pinching or stressing of the disk due to overtight clamping or fit in the mirror cell, and flexure of an inadequately supported disk under its own weight. Thermal effects can easily, and frequently do, bring about a miraculous transformation of a very good mirror into one for which there are no words in polite society; unfortunately it never works this alchemy in reverse! If afflicted with this malady, there is nothing for it but to pack up for the time being while thermal relaxation takes its course or, perhaps, to pass the time with some undemanding low-power sightseeing. One can, however, take common-sense precautions to avoid those recipes which create the problem in the first place, the two worst and commonest being indoor storage at, say, 20–25 °C of an instrument that may be called into play at a moment's notice outdoors at 5 °C or below, and inadequate ventilation and other provisions for temperature stabilisation in small observatories having full exposure to the noonday sun – heating one's telescope to perhaps 40 °C is not a good preparative for high-class images a few hours later!

Mechanical distortion, whether due to pinching or to lack of adequate support of the disk, is essentially a question of mirror-cell design and management, which are dealt with extensively in the large literature of telescope making. There are two basic principles which cannot be overemphasised. Firstly, positive clamping of a mirror in its cell will almost always impair good figure and should be avoided. Secondly, gravitational flexure of a disk of thickness T scales as D^4/T^2, so increasing rapidly with aperture D even for a constant thickness-to-diameter ratio (T/D). The immediate consequence of this last point is that the requirements for adequate mirror support grow rapidly with size of disk, from three-point support which may suffice for full-thickness mirrors up to 10 or even 12 inches diameter, to 18 or 27-point which is necessary for virtually all mirrors of 20 inches and above. The current fashion for lightweight, thin paraboloids is very much more demanding in this respect and it is unlikely, for instance, that a 10-inch of 1-inch thickness will attain the levels of performance referred to here if carried on anything less than a nine-point support system.

Such optical woes are emphasised in this chapter because reflectors are very much more vulnerable to these conditions than refractors, as noted earlier. The conclusion does not follow, however, that Newtonians are inferior to refractors in all the most challenging fields of double-star observation. On the contrary, all the causes of temporary distortion or misalignment of mirrors are avoidable and a good Newtonian well managed will reach the Dawes limit just as well as any refractor.

How to See the Diffraction Limit of any Telescope

Seeing is in some respect an art, which must be learnt
(William Herschel, 1782)

The Airy diffraction pattern is not easy to observe astronomically in its full and perfect glory – practically, never in anything other than a small telescope (less than about 5 or 6 inches in aperture, the comments below referring primarily to larger instruments) under virtually perfect seeing conditions. Otherwise the best one can hope for is a partial, flickering view which it may take long experience as a telescope user to recognise as "diffraction" rather than seeing blur: it took this author over 20 years with the same 12.5-inch mirror. The rings, in particular, are incredibly sensitive to atmospheric distortion, incomparably more so than the diffraction disk itself, and simply vanish without legible trace in Newtonians of typical amateur size, the moment the seeing falls below I or II (Antoniadi). It is therefore of great value to have a means of displaying these and related effects at the level of the telescope's limiting resolution much more clearly, and so to train the eye to see structure at this level.

The first stage is to learn just what the resolution limit of one's telescope actually looks like, just how tiny this really is, how very much smaller than the usual star-image as seen on 90% of nights. It is very easy to go on using a telescope for years, especially if only using powers up to 20 or 25 per inch of aperture, firmly under the impression that the "splodge" one sees a star as at best focus on typical nights is the diffraction disk and that, even if not, there will be no finer level of structure visible in the image. This is wrong even as a rough approximation, but may be a difficult lesson to unlearn and require a change of observing habits. The agitated "fried egg" which one sees in apertures over 6 inches on all except the very finest nights is nothing whatever to do with the true diffraction image, either as to size or structure. Nevertheless, on all except the worst nights, the true limiting-resolution star disk *is* visible, buried in the heart of the obvious image,

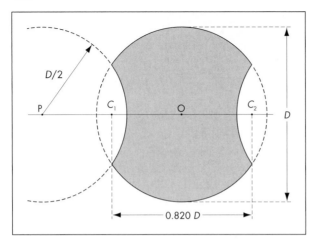

Figure 11.5. The Rayleigh limit aperture mask.

quite accessible (at least in "flashes") to a trained and sufficiently agile eye, perhaps a factor of five smaller than the "splodge". However, no amount of general stargazing will bring about this training of the eye, for which specific exercises are required.

A great aid to this first step of adjusting the eye to the scale of the true diffraction image is a simple aperture mask, of the form shown in Figure 11.5, cut from a sheet of any stiff, opaque material and placed over the aperture (diameter D) of the telescope. Each of the segments symmetrically cut out of the mask is bounded by a circular arc of diameter D struck from a centre P where OP = 0.820 D. A fabrication accuracy of $\pm$ 1/16 inch is perfectly adequate.

With this applied to the telescope, one has a Michelson stellar interferometer specifically designed to produce interference fringes having a spacing exactly equal to the Rayleigh diffraction limit $1.22\lambda/D$ for that telescope. Observe a first magnitude star (*not* a close double!) with this at a power of at least 40 per inch of aperture (40D), focussing carefully. This time, it is not necessary to wait for a night of first-class seeing, as the interference fringes "punch through the seeing" to an extraordinary degree, a surprising and rather curious fact commented on by many users of the interferometer since Michelson himself in 1891. What you will see is an enlarged and elongated diffraction disk divided into extremely fine bright fringes, perhaps as many as 10 or 11 in all (see Figure 11.6). Unless you have done something like this before, you will probably be surprised at how small this scale of image structure is: in all probability a lot smaller than the star images

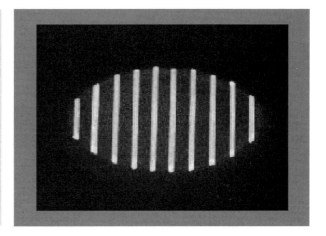

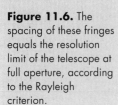

Figure 11.6. The spacing of these fringes equals the resolution limit of the telescope at full aperture, according to the Rayleigh criterion.

usually seen in the same telescope. The magnification required to separate these fringes clearly will depend on your visual acuity and this observation provides an interesting opportunity to test the question of so-called "resolving magnification". The majority of observers will almost certainly find that the commonly alleged figure of $13D$ to $15D$ is hopelessly inadequate and some may need $50D$ or more.

Having accustomed the eye to the appropriate scale of image structure, the next stage is to become thoroughly familiar with the Airy diffraction pattern itself. This is made much easier if the pattern is enlarged *relative to the scale of the seeing* by use of a series of circular aperture stops reducing the telescope's entry pupil to $D/4$, $D/2$, and $3D/4$. It is advisable when doing this with any reflector having a central obstruction to make both the $D/4$ and $D/2$ stops *off*-axis in order to keep vignetting by that obstruction to a minimum. On a night of seeing I or II (Antoniadi) focus the telescope on a second or third magnitude star with a power of at least $50D$ (the author's standard working power for this type of observation is $66D = \times 825$) and keep this same magnification on throughout, while examining the image successively with apertures of $D/4$, $D/2$, $3D/4$ and D. If the telescope is of good quality and properly collimated, you should have no difficulty at all in seeing a nearly perfect "textbook" Airy pattern with the smallest stop: a big, round central disk (not in the least point-like at this power of 200 or more per inch of aperture used), sharply defined, and surrounded by several concentric diffraction rings, extremely fine even on this power, nicely circular and separated by perfectly dark sky.

On running successively through the larger apertures $D/2$, $3D/4$ and D this Airy pattern will shrink dra-
matically and, unless the seeing and the collimation of
the telescope are perfect, it will also suffer a progressive
deterioration. The result on full aperture is unlikely to
bear much resemblance to the ideal image shown by
$D/4$, even ignoring the difference of scale, partly due to
the much greater sensitivity of the larger aperture to
atmospherics and "seeing", and partly to the almost
inevitable residual coma arising from incomplete collimation. Note that equation (11.1) implies that coma at
full aperture D will be 16 times that at $D/4$ for the same
offset θ, so that an asymmetry like that shown in
Figure 11.2, or worse, will now make its appearance
even where none was visible at $D/4$. Nevertheless, if the
night is sufficiently fine, it should be possible with per-
sistence to recognise some trace of the pattern of disk
and rings even on full aperture. Now is the moment to
return to the business of "hyperfine" collimation discussed earlier, completion of which should result in a
perfectly round Airy disk, at least, even though the
rings at full aperture are unlikely ever to be as clean as
those seen at $D/4$. The telescope will resolve to the
Dawes limit *if and only if* this state is achieved; if the
Airy disk absolutely refuses to come round as a button
the instrument is defective and consideration will need
to be given to the possible causes of image distortion
discussed in the previous section or, in the worst-case
scenario, to the imperfections of the main mirror itself.

The final stage of this ocular training programme is
to learn to cope with the seeing on more typical nights
when the diffraction rings will be so fragmented and
perpetually on the jitter as to be completely unrecog-
nisable. Here I refer to seeing down to about III
(Antoniadi), the worst at which high-resolution astronomy is possible. But it is not in the end the rings with
which we are primarily concerned and the emphasis on
them here has been purely for their great sensitivity as
a diagnostic tool, for identifying and curing removable
coma in the telescope. The real image is the *disk* and
the fundamental point about that is that it is often still
there even on second-rate nights when the outer envelope of the seeing blur may reach several times the
Dawes limit. Though then quite invisible to an observer
not specifically trained to work at the diffraction limit,
the Airy disk will time and again reveal itself to a
trained eye as an intense nucleus buried in the heart of
that seeing blur. The object of the exercises suggested

in this section is that it should now be possible, with some further practice on these more typical nights, to do what the untrained eye never could – to pick out the true disk and ignore the atmospheric "noise".

This last stage is perhaps the most difficult, though it should not present great problems if the earlier exercises have been successfully completed, and the requirement now is practice on nights of less than perfect seeing: practice, practice, and more practice. In fact these ocular gymnastics soon become quite easy and instinctive. It is probably in part the lack of such training and consequent failure to distinguish the seeing blur (the gross image outline) from the still visible Airy nucleus which is responsible for the persistent myth that seeing limits ground-level resolution to $1''$ at best, and is certainly the origin of some of the more spectacularly absurd figures one sees quoted for alleged image size. This author's experience of typical conditions at a very typical lowland site may be of some interest in this context: using a 12.5-inch Newtonian at 400 feet elevation (130 m) in central England, an equal $\frac{3}{4}''$ pair (such as η CrB in May 2000) is steadily separated by a clear space of dark sky at x238 in seeing of only III–II (Antoniadi), while PA measures of pairs at $1.8''$ and below are frequently within $2°$ or so of subsequently verified definitive values even when the seeing is III (e.g. Σ 138 Psc January 2000 and ξ UMa, April 2000, both at ×238). These observations prove that the mean angular size even of the gross outline of the image as seen under such very middling conditions is no more than about $0.6''$, in the centre of which the smaller Airy nucleus is still fitfully visible. When the seeing improves to I or II this accuracy of PA measures extends down to pairs at $1''$ or even slightly below, and this is using the most primitive of home-made micrometers on an undriven alt-azimuth telescope.

When described minutely like this, the business of fine-tuning the capabilities of instrument and observer is perhaps likely to appear a rather arduous road. In fact, this could scarcely be further from the truth, as the training of the eye is essentially once-and-for-all, while one soon drops into a virtually unconscious habit of the collimation procedures described earlier, which then take merely a few minutes at the start of each observing session. While it must be emphasised in the strongest terms that, as Herschel put it, "you must not expect to *see at sight*", there is no obvious reason why a new observer, starting from scratch and following the programme outlined in this section, should have any difficulty in attaining a fully trained eye within a few months of commencing observations. I believe the value of the approach outlined in this section lies entirely in making that possible – it is certainly not necessary for the process to take the 20 years it took this author (with the same telescope) in the absence of any such detailed guidance!

Achievable Results

So, what sort of performance and results can one expect from a fairly typical amateur reflecting telescope, say of 6–12 inches or so aperture and of good optical quality? Without the small investment of trouble in adjustment of the instrument and training of the eye outlined in the preceding paragraphs, the field of wide doubles is open to the observer, that is to say pairs from 1–2″ upwards. Diffraction-limited performance will not be attained by a substantial margin and such an observer will probably consider resolution of a 1″ pair something of a triumph, while subarcsecond doubles remain an unattainable holy grail. Much rewarding observation can be done in this rather undemanding way but that ingredient which gives double-star astronomy its deepest fascination will be largely lacking: *motion*. Very few of these wider pairs have orbital periods of less than centuries so the observer limited to this type of observation is largely condemned to studying binaries as static showpieces, missing out thereby on the grandest gravitational ballet in the whole of celestial dynamics. Adding the dimension of time, and being able to watch these majestic systems actually in action, adds incomparably to the interest of the observations.

The representative selection of this author's observations quoted below illustrate what can be done with very ordinary amateur equipment in this dynamic, subarcsecond domain, given the attention to preliminaries described above. The instrument used is the 12.5-inch (0.32-m) Newtonian referred to earlier and shown in Figures 11.7 and 11.8. It has a plate glass primary mirror figured by George Calver in 1908 which, as discussed in the first section, was deliberately left undercorrected by its maker, with the residual

Figure 11.7. The 12.5-inch reflector (Peter Seiden).

spherical aberration consequently tending to give rise to diffraction rings of largely enhanced intensity. While, as pointed out earlier, the effect of this is to make such a reflector no match for a good refractor on very unequal pairs below about 1.5″, the spurious disk remains at the ideal Airy size or even slightly smaller, so equal ($\Delta m \leq 1$) close pairs can be resolved at least as

Figure 11.8. Eyepiece end of the 12.5-inch (Christopher Taylor).

Observing and Measuring Visual Double Stars

well as in a refractor of the same aperture. Accordingly, the results quoted are all for binaries whose components do not differ by much more than 1 magnitude in light.

Lest the reader imagines that successful observation of subarcsecond binaries requires an expensive professionally constructed instrument equipped with the latest hi-tech. conveniences, or that the author has enjoyed such advantages in making the observations reported here, a brief description of the 12.5-inch will serve as a useful counterexample. The telescope was entirely amateur-built some 60 years ago and, although standing about 9 feet (2.7 m) tall, it has never been housed in any form of building or weatherproof cover. One result of this is that while the mechanical structure of the instrument stands permanently on a concrete foundation at a good observing site in the author's garden, the entire optical system must be stored indoors and mounted anew in its various cells, etc. at the beginning of each observing session. This, of course, means that full collimation of the system is an unavoidable necessity every time it is used – the telescope simply would not work otherwise. Thanks to intelligent design, however, this entire optical assembly and collimation routine only takes five minutes or so each evening: on the general view, Figure 11.7, note:

(a) that all optics are mounted externally, and very easily accessible, on the "tube" which in reality is nothing more than a box-girder for rigidity;

(b) the linkage rods running from the inner corners of the fully adjustable main mirror cell, up the length of the tube, to the eyepiece assembly at the top; these terminate in the collimation control knobs which can be seen at the lower corners of the eyepiece turret housing in Figure 11.8 and make fine adjustment of squaring-on of the main mirror while simultaneously looking through the eyepiece or drawtube a very quick and painless affair.

The instrument weighs about 1500 pounds (680 kilograms) and is an alt-azimuth, lacking not only (therefore) setting circles or clock drive but even any form of manual slow-motion controls. It is true that the 12.5 inch moves very smoothly on its bearings and is extremely stable but it remains something of an acquired skill to follow the diurnal motion at high power simply by pulling directly on a handle at the top of the tube, to say nothing of taking PA measures of close double stars! The full force of this remark will perhaps be appreciated when it is borne in mind that an equatorial star takes rather less than ten seconds to cross the full field of view at the power most commonly used for "subarcseconders", and that the observer, perched precariously on a step-ladder some considerable height above the ground, must perform this manual tracking continuously, coordinating hand and eye to a precision of a few tens of arcseconds, at the same time leaving the mind free to concentrate on what is seen in the eyepiece. This is observing in the classic style of William Herschel, far removed from the digital conveniences of the twenty-first century.

This telescope has a primary focal ratio of 7.04, with a central obstruction equal to 16.3% of the aperture diameter. Optical quality is such that the author's standard working power for all subarcsecond double stars is ×825, at which single stars appear "round as a button" whenever the seeing is II–III (Antoniadi) or better, and the instrument would comfortably bear magnifications even higher were it then possible to manage its alt-azimuth motions sufficiently well. It is clear from the observations that the smallest double-star separation detectable with the 12.5-inch (see below) is, even so, limited by magnification, *not* by definition and image quality. Statistical analysis of accumulated observations of equal bright pairs at 0.4–0.9″ shows that the apparent star-disk diameter of a fifth or sixth magnitude star at ×825 in good seeing is 0.311 ± 0.037″; this observed size of spurious disk is only 37% of that of the full theoretical Airy disk (out to first zero) and agrees exactly, after scaling for aperture, with the result independently determined for a 4 inch (0.102 m) refractor also used for double-star observation. This last comparison shows that the image definition of a Newtonian can be not only as good as that of a refractor of the same aperture but, after scaling, can match that of a far smaller refractor, a much more severe test. It must be emphasised once again, however, that such quality of imaging can *only* be expected of a Newtonian even at this f-ratio after full and accurate collimation, as detailed earlier.

The double-star results actually achieved with this 12.5-inch Newtonian are best represented by Table 11.1 which lists the typical appearance at ×825 of bright, approximately equal pairs at successively smaller separations, in seeing II–III (Antoniadi) or better. Listed here are only those categories of target which can in

Table 11.1.

Separation "	Typical appearance of star disks
0.4–0.5 or so	Two completely separate disks parted by a small gap, persistently split in good seeing, e.g. λ Cas 1995.03 (0″43), φ And 1995.80 (0″48), β Del 1998.72 (0″50), ω Leo 1996.26 (0″52), 72 Peg. 1994.78 (0″53)
0.35–0.36	Two distinct disks in contact (tangent), occasionally *just* separating in good seeing, e.g. β Del. 1996.87 (0″35–6)
0.33–0.34	Disks now slightly overlapping, giving "figure 8", heavily notched but *not* separating, e.g. δ Equ. 1995.79 (0″33–4), α Com 1996.46 (0″33)
0.29–0.32	Very elongated single image ("rod"), occasionally just notched at best moments; an easy elongater, the disk elongation quite obvious in even moderate seeing, e.g. δ OriAa (Hei 42) 1998.11 (0″31).
0.24–0.28	A single oval disk ("olive"), the elongation still quite pronounced although noticeably less than in the last case, no hint of a notch now, e.g. β Del 1995.85 (0″28), α Comae 1997.35 (0″26), A 1377 Dra 1997.80 (0″25), γ Per (WRH 29 Aa) 1996.88–1997.19 (0″24)
0.21–0.23	Slightly oval disk, elongation small but still quite sufficient to read PA confidently at best moments; now becoming noticeably more difficult, the difference between 0.21" and 0.24", very obvious to the eye, e.g. κ Peg. 1996.88 (0″21)
0.17–0.20	Very slightly oval disk, the elongation very small but in the best seeing absolutely definite, especially by comparison with a neighbouring single star as a "control"; now becoming difficult to estimate PA confidently, the detection of elongation nearing the limit for ×825, e.g. ζ Sge (AGC 11) 1996.77 (0″19) which was appreciably easier than α Com. 1998.41 (0″175) – the current limit for positive detection of a double star with the 12.5-inch at this power.
Somewhere at, or above, 0.13	Beyond the limit for reliable detection at ×825, the star disk *not* clearly distinct from that of a neighbouring single star even in very good seeing, e.g. κ UMa (A 1585) 2000.23 (0″13)

any sense be considered seriously testing of the telescope's capabilities, all wider pairs always appearing on any good night as two well-separated stars divided by a large space of completely dark sky.

There is no doubt that such performance claims run heavily counter to the perceptions of the large majority of telescope users who are, perhaps, too undemanding of their instruments. To any reader inclined to be sceptical of the above results I would point out that the author had been using this same telescope on double stars and other "high resolution" targets for more than 25 years before the *observations themselves* forced the possibility of such subarcsecond performance on the attention of a mind not predisposed to expect it;

further that all such double-star observations are made essentially "blind", the observer having no prior information on "expected" PA, and only a rough figure for separation, on going to the eyepiece. So the relentless internal consistency of the observations with respect to separation, and their close individual agreement in virtually all cases with definitive values of PA subsequently consulted as an objective verification, are more than sufficient to establish the objective validity of these results. In the entire set of observations of pairs below 0.5″ there are only two or three cases of clear contradiction with this post-observational check, none of which were in good seeing. These very few failures are, moreover, offset by a number of other instances of apparent contradiction where more authoritative data subsequently obtained have proven that the observations were correct and that it was the published information available at the time which was in error, this having occurred for λ Cas, α Com, γ Per, δ Ori Aa and κ UMa.

Such results should really occasion no surprise as they are actually in precise agreement with the Dawes limit (0.365″ here) as can be seen from the first three classes in Table 11.1, as well as agreeing pretty closely with what would be expected from the previously quoted size of star disk determined quite independently from observations of much wider pairs. All of this is, in fact, entirely in line with the mainstream of historical experience in this field, from Herschel who founded subarcsecond double-star astronomy in the early 1780s with a 6.2-inch mirror (0.157 m), right down to the *Hipparcos* satellite observatory which made accurate measures of pairs down at least to 0.13″ in the early 1990s with a 0.29 m. mirror – rather smaller than that used by the author, although admittedly having the huge advantage of perfect seeing! The limit on detectable separation, for instance, in Table 11.1, at 0.48 of the Dawes limit, is closely comparable with the average for the closest class of new discoveries made by S.W. Burnham with a 6-inch aperture, i.e. 0.53 × the Dawes limit.

The author's observations therefore establish conclusively that double-star elongation of reasonably equal pairs is reliably detectable in a 12.5-inch mirror at ×825 down to a limit somewhere about 0.17″, as with α Com in May 1998. All such pairs down to 0.24″ inclusive are easy elongaters in good seeing, only the last two classes in Table 11.1 really presenting any significant difficulty under the best conditions. What is perhaps most remarkable about such observations is the extraordinary sensitivity of the shape of blended or partially resolved double-star images to really minute changes in separation: in the 12.5-inch at ×825, a change of only 20–30 mas is quite appreciable to the eye in pairs from 0.4″ downwards, while an increase of 60-80 mas is sufficient to transform the appearance of a pair totally, from "olive" to "disks tangent", as in the case of β Del 1995–1996. It is amazing but true that a ground-based amateur telescope of unremarkable aperture and positively primitive lack of sophistication, used visually in the time-honoured fashion, can and does reveal clearly angular displacements smaller than any detail actually resolved by the *Hubble Space Telescope*.

Access to this subarcsecond domain opens the door on a dynamic world of binary-star astronomy usually considered the exclusive preserve of the professional using powerful instruments equipped with the latest technology and sophisticated methods such as speckle interferometry. Indeed, several of the pairs mentioned above have been used in recent years as test objects for evaluating the performance of adaptive optics systems on professional telescopes of 1.5-m aperture and above, while the entries in the third CHARA catalogue show that all are favourite targets of the speckle inter-ferometrists. It is one of the better-kept secrets of observational astronomy that it is nonetheless perfectly possible, with care and determination, to follow many of these systems' orbital motion visually with an amateur telescope of only slightly larger than average aperture, which means, almost necessarily, a reflector. This should not be a surprise to anyone: almost all of these binaries were, in fact, discovered in just this fashion, using very much this range of apertures, by e.g. Otto Struve with the Pulkova 15-inch, Burnham with 6 and 9.4-inch instruments, etc.

Among the author's more memorable experiences with the 12.5-inch telescope are several concerning some of the most legendary of the short-period visual binaries. δ Equ (OΣ 535), perhaps the most famous of all such systems, was long the holder of the record for the shortest period of all visual binaries, at 5.7 years. This pair is actually quite easy on a good night in the 12.5-inch when at its widest as in 1995, appearing then as an absolutely unambiguous figure 8, only just failing to separate completely. The orbital motion is phenom-

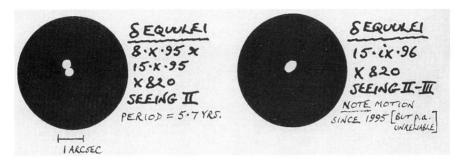

δ EQUULEI
8·X·95 x
15·X·95
X 820
SEEING II
PERIOD = 5·7 YRS.

δ EQUULEI
15·iX·96
X 820
SEEING II-III
NOTE MOTION
SINCE 1995 [BUT p.a. UNRELIABLE]

I ARCSEC

Figure 11.9.
Observations of the pair δ Equ with the 12.5-inch reflector.

enally rapid, a total transformation in the appearance of the star occurring in a year or less, as the author witnessed in 1995 and 1996 – see Figure 11.9. This motion is actually so rapid that, if caught in very good seeing at the critical moment in the orbit, a change plainly perceptible at the eyepiece of the 12.5-inch will occur in only seven or eight *weeks*, δ Equ having crossed an entire class in Table 11.1.

β Del (β 151, period 26.7 years) is another pair whose orbital advance in a single year is plainly visible in the 12.5-inch reflector even without quantitative measurement, its steady year-by-year opening out and rotation in PA having been conspicuous in that telescope in the years 1995–1998. This was first noted on 13 November 1996, the entry for which in the author's observations (obs.) book reads "β Delphini ×820 showing an immediately obvious 'rod'/'figure 8'; on further scrutiny, several times glimpsed two distinct stars just touching, i.e. this pair now much easier than a year ago … PA constantly and easily legible at 330–335°." This was a rough "by eye" estimate only, not a measurement, but very noticeably larger than it had been twelve months earlier, the seeing only fair at III–II. The definitive position at the time of this observation was subsequently found to be (0.35–0.36″, 323°). See Figure 11.10 (overleaf).

Other similar cases have been α Com (Σ1728, period 25.9 years); γ Per (WRH 29 Aa, period 14.7 years) a beautiful system which is the brightest visual binary in the heavens also to be an eclipsing variable[11], the double-star observations of which have been mostly by speckle interferometry on 3 to 4-metre class telescopes; and κ Peg (β 989, period 11.6 years). The 12.5-inch followed the inward march of Σ1728 over the late 1990s, beginning with "figure 8" at 0.33″ in 1996, all the way down to "elongation v. slight but perfectly definite" at

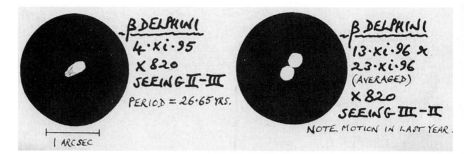

Figure 11.10.
Observations of the pair β Del with the 12.5 inch reflector.

0.175″ in 1998, the smallest separation so far detected with this telescope. The annual change in this star was quite apparent at each of these three observing seasons and, although it was much more difficult in 1998 than it had been a year earlier at 0.26″, even the limiting elongation to which it was followed was quite unmistakeable – "like a dumpy egg" – by repeated comparison with the absolutely round disk of Arcturus, then at the same zenith distance. (Given that 0.33–0.34″ appears as "figure 8", this is in fact exactly what one should expect of the same pair at 0.175″, as can easily be seen from scale drawings of spurious disks overlapping to the appropriate degrees.)

Some Advice

If such are the results achievable with the decidedly primitive amateur-built telescope described earlier, it must follow that similar performance is within reach of virtually any Newtonian having a good mirror at f/5–6 or longer, adequately supported on a mounting of sufficient stability and rigidity. Those further refinements which the author's instrument so conspicuously lacks – permanently mounted optics of modern low-expansion glass in a telescope having a clock drive or at least good manual slow motions – will, of course, make this easier but are not indispensable. The real essentials for such subarcsecond performance are listed here, together with some general points of advice on the conduct of this type of double-star observation:

(i)   While any good instrument is worth giving a fair trial on subarcseconders, it is unlikely, in the case of reflectors especially, that a system having a primary f-ratio of less than 5 will achieve the level of performance described above, even if claimed to be "diffraction limited" (a decidedly loose phrase): equation (11.3) makes it clear that collimation tolerances for critical imaging quality become almost impossibly tight at $F < 5$, in addition to which these deeper curves of the main mirror are more difficult for the optician to control by most of the methods of figuring and testing still in use, so that such "fast" paraboloids are rarely as good as the best of longer focus. In general it is clear for reflectors that the longer the focus the better, within reason; even $F = 12$ or 15 would certainly not be excessive here.

(ii) It goes without saying that such extremes of imaging performance can only be expected of good optics, of course, but it would be a mistake to suppose that the author's 12.5-inch is wildly exceptional in this respect. Calver was undoubtedly a master optician but he was working with both materials and methods which made his job decidedly more difficult than that of his modern successor; there must be many more recent mirrors in amateur hands which are just as good as this 1908 glass. It is probably true that any paraboloid as good as, or perhaps a little better than, the Rayleigh quarter-wave criterion will deliver the sort of results described here, if well managed and satisfying the other necessary conditions. Remember, however, that the Rayleigh criterion means that the *extreme* distortion peak-to-valley of the *wavefront* must not exceed one quarter of the relevant wavelength of light used; a phrase such as "a one-tenth wave mirror" may, in extremis (and often does!), mean that the mean deviation of the glass from perfect figure does not exceed one tenth of a test wavelength (usually He–Ne laser at 6328Å) which is itself considerably larger than the 5100–5300Å value relevant to visual observation. In such terms, a surface only just satisfying the Rayleigh criterion would be described as "one-thirteenth wave", so beware ambiguous descriptions of optical quality from telescope retailers, manufacturers and others!

(iii) On the needlessly controversial subject of magnification, the only rule is that there are no rules, and any attempt to set hard and fast limits to what may be used on a given aperture is merely an arbitrary and unhelpful constraint hampering

the realisation of the telescope's uttermost capabilities. The wise observer will give full play to the
instrument's whole range of powers without prejudice and finally settle on that magnification
which best reveals the details sought, irrespective
of whether that also yields the crispest, aestheti-
cally most satisfying image. The last is a merely
cosmetic consideration. As to high, or even very
high, powers – say from 40 per aperture-inch
upwards – be neither obsessed with, nor afraid of
them. It should be pointed out that the "resolving
magnification" is the theoretical *minimum* for
visibility of small detail, not a maximum; oft-
repeated attempts to set this as an upper limit to
useful magnification, taking 1′ as the smallest
detail resolvable by the eye and Dawes' or
Rayleigh's limits as the smallest that one may be
attempting to see with the telescope, are fallacious
on all counts: visual acuity varies hugely from one
individual to another but the typical night-time
resolution of a normal eye is $2\frac{1}{2}$ to 3′, not 1, while
Table 11.1 of subarcsecond double-star appear-
ances shows that we may very well be in quest of
detail as small as 0.5 times the Dawes limit, to
magnify which up to comfortable visibility therefore requires a power of at least 65 per aperture-
inch, a figure itself not in any sense an *upper*
limit. This is quite in line not only with the
author's experience with the 12.5-inch mirror
(×65.8 per inch) and Jerry Spevak's with the 70-
mm objective (×72.4 per inch, see Chapter 10) but
also that of most observers of such close visual
pairs. You may be able to reach these subarcsecond limits at substantially lower magnifications
but I shall be surprised!

(iv) A vital corollary of the last point is that the whole
mechanical construction of the telescope must be
such that both its rigidity and smoothness of
movement are able to handle the high
magnifications necessary. This is a rather
demanding requirement, which in larger
apertures is virtually certain to be incompatible
with the lightweight construction favoured for
portable telescopes, many of which are hugely
under-engineered in this respect. For a reflector
over about 6 inches aperture, a permanently
mounted instrument is certainly better than a
portable for this class of observing and it is

evident from this consideration and point (i) that the popular f/4.5 Dobsonian of large aperture is just about the worst possible choice here. Such telescopes are not the tools of high-resolution astronomy.

(v) Full and thorough collimation of a reflector's optics as frequently as may be needed to maintain their precise alignment is an absolute essential, as discussed earlier. Equation (11.2) now makes it very obvious that the smallest errors of squaring-on at the arcminute level will be quite sufficient for coma to swamp many of the finer features in Table 11.1.

(vi) The quality of the seeing is of vital importance. Don't waste time attempting to observe subarcseconders when the Airy disks of these stars are not visible (say, seeing III Antoniadi or worse).

(vii) These pairs should only be observed when at a large elevation above the horizon, preferably within about 1 hour of meridian passage, and certainly not when below about 35°. Below 40° elevation, elongation of star disks due to atmospheric spectrum becomes increasingly evident and the seeing steadily deteriorates due to the lengthening visual ray within the turbulent atmosphere. Resist the temptation to try for subarcseconders which never rise above these elevations in your sky–the results will only be gibberish.

(viii) This sort of observing does not require phenomenal eyesight; the author is slightly short-sighted and certainly of only average visual acuity even when corrected for myopia. What it does emphatically require is a mental receptiveness to every nuance of what is seen, a power of concentration which devours to the last drop what the eye has to offer. This ability to *use* one's eyes takes training and practice, of which something has already been said earlier. It is remarkable how widely telescope users differ in this respect, even among active observers, but fancy equipment is *no* substitute here for essential observing skills. In training the eye to this activity it makes obvious sense not to be too ambitious at first but to start with pairs at several times the Dawes limit and then work steadily downwards. The furthest fringes of subarcsecond double-star observing are undoubtedly an extreme sport, a sort of "athletics for the eyes", which demands fitness as with any such activity. Illness, tiredness or significant alcohol intake are all quite incompatible with peak performance, which depends as much on the observer as it does on the instrument.

Spectacle wearers must, necessarily, abandon their glasses for this work, as the high magnifications used require eyepieces whose eye-relief is much too small to accommodate them. This is no problem whatever to those suffering only from pure long- or short-sight as simple re-focus of the telescope takes care of all, but astigmatism is a more serious matter. Uncorrected, this will cause spurious elongation of star disks with obviously undesirable consequences, so the astigmatic observer who would pursue this game must resort either to contact lenses or to a tight-fitting eyepiece cap carrying the appropriate corrective glass (e.g. old spectacle lens or a piece cut centrally from one).

(ix) Unequal close pairs are *much* more difficult than equal pairs at the same separation, especially in reflectors generating accentuated diffraction rings, in which an inequality of even 1 magnitude may cause considerable difficulty in the clear sighting of a companion anywhere near the first ring, and a magnitude disparity of 2 or only a little more makes it practically invisible. Most of the remarks above concern approximately equal pairs (magnitude difference less than 1, say) and it makes sense to begin with these on first setting out to crack subarcseconders. An illustrative example here is Albireo, the bright component of which is itself a very close double (MCA55) having a brightness inequality of about 2 magnitudes: at 0.38″ this is very much more difficult in the author's 12.5-inch (e.g. obs. 1996.80) than an equal pair such as δ Equ at 0.33″, probably, in fact, as difficult as any pair successfully observed with that telescope. The effects of seeing and of use of different optical systems on the detectability of these unequal pairs is altogether a more complex affair than the corresponding questions for equal doubles, and their observation consequently yields much less reproducible results.

(x) For all really doubtful or difficult cases, Herschel's advice could not be bettered: while leaving the eyepiece and focus untouched alternate in quick succession between views of the target double and of a nearby single star at about the same altitude, so using the roundness of the latter as a "control" or comparison for the observed disk shape of the double. If the comparison star shows any significant elongation, the entire observation should be rejected.

(xi) Lastly, we come to perhaps the most important point of all for any observations which may with any justification be challenged or doubted, in which category should probably be included all alleged sightings of pairs equal or unequal, separated by less than twice the Dawes limit for the instrument used. As a matter of elementary scientific method it is *essential* that the observer has some independent means of checking each observation and so proving its validity to the sceptic (quite possibly the observer themselves). This requires that the observation is always made "blind" with respect to some observable parameter of the pair, the observer having deliberately gone to the eyepiece *not* knowing everything about the current appearance of the target, so that the only possible source of knowledge of the parameter is the observation itself. The observed value can then, post-obs., be checked against the "correct" or expected value as an objective criterion of verification (OCV). The most obvious choice of OCV is the position angle. Thus, and only thus, can observer prejudice, the phenomenon common in some less rigorous visual astronomy of "seeing what you expect to see", be eliminated and these extremes of double-star observation be securely founded on objective detection of the chosen targets. (This is flatly contrary to the (bad) advice given in some handbooks but it must be recognised that questionable observations made in the absence of any OCV, or where none is possible (e.g. as in claims to have seen the central star of the Ring Nebula M57 with small telescopes), are quite meaningless.) If in any doubt about PA at the first observation of a difficult pair where the seeing is less than ideal, do not check the value then but re-observe the target on better nights until confident of the result, and only then consult the OCV.

To conclude, enough has surely now been said to make a powerful case for the reflecting telescope as fully the equal of the refractor, aperture for aperture, in at least some of the most demanding classes of double-star observation. The author hopes that this may be an encouragement to users of good reflectors to venture into a deeply fascinating field of observation from which the speculum has too often been unjustifiably excluded by false preconceptions of the superiority of the lens.

References

1   Conrady, A.E., 1919, *Mon. Not. R.. Astron. Soc.*, **79**, 582.
2   Couteau, P., 1981, *Observing Visual Double Stars*, MIT Press, p. 32.
3   Born, M. and Wolf, E., 1999, *Principles of Optics*, 7th edn, Cambridge University Press, p. 424.
4   Born, M. and Wolf, E., Chapter IX, esp. pp. 520–2 and refs. given there, esp. Wolf 1951, pp. 106–8.
5   Ceravolo, P., Dickinson, T. and George, D., March 1992, *Sky & Telescope*, Optical Quality in Telescopes, 253.
6   Taylor, H.D., 1983, *The Adjustment and Testing of Telescope Objectives,* 5th edn, Adam Hilger, Bristol, pp. 41–4, in which the ray, or geometrical optics, picture is used.
7   Dimitroff, G. and Baker, J.G., 1947, *Telescopes and Accessories,* Harvard University Press, pp. 42–3.
8   Berrevoets, C., 2000, *Aberrator*, Version 2.52 (I am indebted to Cor Berrevoets for the use of this excellent programme to generate Figures 11.1–11.4, and to David Randell for drawing my attention to *Aberrator* and to its availability as internet freeware at <u>http://aberrator.</u> <u>astronomy.net</u> (the version now available is Version 3.0 – Editor).
9   Bell, L., 1922, *The Telescope*, New York, p. 95.
10 Sidgwick, J.B., 1979, *Amateur Astronomer's Handbook,* 4th edn, Pelham, London, Chapter 12,
11 Griffin, R.F., June 1991, *Sky & Telescope*, Gamma Persei Eclipsed, 598.

Further Reading

Hysom, E. J., 1973, *J. Brit. Astron. Assoc.*, **83**, 246–8.

Chapter 12

Simple Techniques of Measurement

Tom Teague

Background

It is not necessary to possess expensive or advanced apparatus in order to begin making accurate measures of double stars. This chapter discusses three different techniques, in ascending order of sophistication: the ring method, the chronometric method, and finally the use of reticle eyepieces. Of these, the ring method is the simplest, requiring in its crudest form nothing more than an ordinary stopwatch with lap facility. By the addition of a crosswire and position angle dial, the observer can begin to measure closer pairs. Even an illuminated reticle eyepiece requires no great financial outlay, and permits observations comparable in accuracy with those achieved using a filar micrometer.

The Ring Micrometer

Invented by the Croatian Jesuit astronomer Roger Boscovich (1711–87), this is an elegant method of measuring differences in right ascension and declination. In its true form, the ring micrometer comprises a flat opaque ring mounted at the focus of the telescope objective. Using a stopwatch, the observer times transits of double stars across the ring. The times at which the components cross the inner and outer peripheries of the ring, together with the declination of the primary component and the known value in arcseconds of the ring diameter, contain all the information necessary to calculate the rectangular coordinates of the pair (i.e. the differences in right ascension and declination separating the two stars), from which it is then possible to derive its polar coordinates (ρ, θ).

It cannot be denied that the mathematical process of reducing the results is somewhat cumbersome, and must have been almost prohibitively tedious in the days of slide rules and logarithm tables, but the advent of modern electronics has banished such difficulties forever. The observer who makes good use of a computer or programmable calculator need not be deterred by the mathematical complexities which are, in any case, more apparent than real.

Commercially made ring micrometers are no longer obtainable, and the construction of a good one is not for the faint-hearted. My own, manufactured by Carl Zeiss Jena, consists of a metal ring mounted on a centrally perforated glass diaphragm which is fitted at the focus of a positive eyepiece. Happily for those who prefer not to undertake their own precision engineering, it is not actually necessary to have a purpose-made ring micrometer. All that is required is an eyepiece having minimal field curvature and an accurately circular field stop. It is the latter which serves as the micrometer. Some modern eyepieces, though of acceptable optical quality, have plastic field stops that may not be truly circular. Select a good-quality eyepiece with a flat field and a metal field stop. It is possible to flatten the field by incorporating a Barlow lens into the optical train.

The first step is to calibrate the eyepiece by determining the radius of its field in arcseconds. A simple method of doing this is to time how many seconds of mean solar time it takes a star of declination δ to drift across the field diametrically, multiplying the result by 7.5205 cos δ. Even the mean of a number of such timings, however, is unlikely to be very accurate, since the observer has no way of being sure that the star has passed through the exact centre of the field of view, as opposed to trailing a chord.

A more reliable calibration method is to use a pair of stars having declinations which have been determined to a high degree of precision. The *Tycho-2* catalogue will yield plenty of suitable candidates. In order to minimise the effects of timing errors, choose stars of relatively high declination, between 60 and 75° north or south of the celestial equator. The difference in dec-

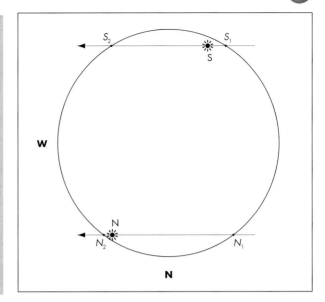

Figure 12.1. Timing the transits of a wide pair of stars to determine the accurate diameter of a field stop or ring.

lination of the two stars should be slightly less than the diameter of the field stop or ring. Their separation in right ascension is less important, but should obviously not be inconveniently large.

The two stars are allowed to drift across the field, so that one star, N, describes a chord near the north edge of the field and the other, S, near the south edge. The times at which each star enters and leaves the field are recorded using a stopwatch (Figure 12.1). A cheap electronic sports watch with lap counter will be found perfectly adequate.

Two angles, X and Y, are required in order to calculate the precise radius of the field stop in arcseconds. Suppose that star N, of known declination δ_N, enters and leaves the field at N_1 and N_2, respectively, and star S, of declination δ_S, enters and leaves at S_1 and S_2. Let $\Delta\delta$ be the difference in declination between the two stars. Then:

$$\tan X = \frac{7.5205(S_2 - S_1)\cos\delta_S + 7.5205(N_2 - N_1)\cos\delta_N}{\Delta\delta}$$

(12.1)

$$\tan Y = \frac{7.5205(S_2 - S_1)\cos\delta_S + 7.5205(N_2 - N_1)\cos\delta_N}{\Delta\delta}$$

(12.2)

from which the radius of the field, R, may be derived as follows:

$$R = \frac{\Delta\delta}{2 \cos X \cos Y}. \tag{12.3}$$

Take the mean of not fewer than 30 transits. For the greatest possible accuracy, allow for the effects of differential refraction (see Chapter 22).

The procedure for measuring a double star is as follows. Set and clamp the telescope just west of the pair to be measured, so that the object's diurnal motion will carry both components, A and B, across the field as far as possible from its centre (Figure 12.2); they should both transit the field near the same (north or south) edge unless they are very widely separated in declination, in which case they may pass on opposite sides of the centre of the field. The importance of ensuring that the stars pass close to the north or south edge is that it minimises the impact of timing errors upon $\Delta\delta$. However, it should not be carried to extremes, as the precise moment of ingress or egress of a star that merely grazes the field edge will eventually become impossible to pinpoint.

The first transit should be used as a "reconnaissance" to determine and record the sequence of appearances and disappearances. On subsequent transits, the observer uses a stopwatch to obtain the times (A_1, A_2, and B_1, B_2) at which each star enters and leaves the field; these times are noted in tabular form, as shown in Table 12.1.

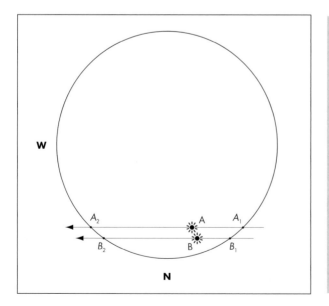

Figure 12.2. Using the eyepiece field stop or ring to measure a pair of stars by transits.

Table 12.1. A specimen observation of Σ I 57 made on 1997, October 27. The three transits are individually numbered in the top row of the table. Also recorded in each column is the portion of the field in which the transit took place (i.e. north or south, as the case may be)

	1 (N field)	2 (N field)		3 (S field)
A_1	0.00	0.00	B_1	0.00
B_1	29.36	30.95	A_1	3.56
B_2	276.77	270.68	A_2	251.80
A_2	279.81	275.06	B_2	281.69

In order to calculate the position angle, θ, and separation, ρ, of the pair, it is first necessary to determine the differences in right ascension, α, and declination, δ, between the two components. The time at which each star transits the centre of the field is given by the mean of the times at which it enters and leaves. Hence the difference, $\Delta\alpha$, in RA between the two stars, A and B, is given by:

$$\Delta\alpha = \frac{B_1 + B_2}{2} - \frac{A_1 + A_2}{2}. \tag{12.4}$$

The result is expressed as a time difference. At a later stage, after we have ascertained the individual declinations of both components, we will be able to convert $\Delta\alpha$ to its great circle equivalent, in seconds of arc.

In order to obtain the difference in declination, $\Delta\delta$, between the two stars, we first need to ascertain the distance, D, in declination between the centre of the field and each of the stars, A and B:

$$D_A = R \cos \gamma_A \tag{12.5}$$
$$D_B = R \cos \gamma_B \tag{12.6}$$

where the angles γ_A and γ_B are given by the following equations:

$$\sin \gamma_A = \frac{7.5205 \cos\delta_A (A_2 - A_1)}{R} \tag{12.7}$$
$$\sin \gamma_B = \frac{7.5205 \cos\delta_B (B_2 - B_1)}{R} \tag{12.8}$$

The difference in declination between the two objects is then given by:

$$\Delta\delta = D_A \pm D_B \tag{12.9}$$

Table 12.2. How to assign a position angle to its correct quadrant. Note that for the purpose of using this table, the sign (+ or –) of $\Delta\delta$ is always taken from a transit carried out in the northern half of the field; otherwise the signs must be reversed.

$\Delta\alpha$	$\Delta\delta$	Quadrant	$\theta =$
+	–	1 (0 – 90)	θ
+	+	2 (90 –180)	$180 - \theta$
–	+	3 (180 – 270)	$180 + \theta$
–	–	4 (270 – 360)	$360 - \theta$

The value of D_B is added to D_A when the stars are on opposite sides of the centre of the field and subtracted from it when, as is more usual, they are on the same side. Note that in the latter case, the sign (positive or negative) of $\Delta\delta$ varies according to whether the north or south portion of the field is used. When both stars pass to the north of the field centre and $\Delta\delta$ is positive, B lies south of A; a negative result indicates the contrary. When both stars pass to the south of the field centre, the rule is reversed.

Since only the declination, δ_A, of the main component, A, is usually known in advance, the declination, δ_B, of the secondary component, B, must initially be given the same value for a first approximation. Once a preliminary value has been derived for $\Delta\delta$, the result is added to or subtracted from δ_A (as the case may be) to obtain a refined value for δ_B, from which $\sin \gamma_B$ and thence $\Delta\delta$ may be recalculated.

We are now in a position to convert $\Delta\alpha$ into arcseconds. To do this, multiply by 15.0411 cos δ, where δ is the mean declination of both stars.

Having thus obtained final values for $\Delta\alpha$ and $\Delta\delta$, we use simple Pythagorean trigonometry to work out the polar coordinates, ρ and θ and:

$$\theta = \tan^{-1}\left(\frac{\Delta\alpha}{\Delta\delta}\right) \qquad (12.10)$$

$$\rho = \sqrt{\Delta\alpha^2 + \Delta\delta^2} \qquad (12.11)$$

When calculating θ, it is necessary to allow for the quadrant in which the companion (B) star lies by applying the appropriate correction, as shown in Table 12.2.

The first transit of the star Σ I 57 recorded in Table 12.1 provides a convenient practical example. We can see that the difference in right ascension, $\Delta\alpha$, is given by (12.1):

$$\frac{(29.36+276.77)}{2} - \frac{(0.00+279.81)}{2} = 13.16 \text{ seconds}$$

Let us now calculate the difference in declination between the two components. The first step is to find the angles γ_A and γ_B. Consulting our catalogue, we find that the declination (2000) of Σ I 57 is +66°.7333 (this refers to the A component). The radius of the ring used to make the observation was 916″. Therefore:

$$\sin \gamma_A = \frac{(7.5205 \times \cos 66.7333 \times (279.81 - 0.00))}{916} = 0.9075$$

from which it follows that γ_A itself must be 65.16. By the same method, we find $\sin \gamma_B$ to be 0.8024, and $\gamma_B = 53.36$ (note that at this stage, in the absence of an accurate figure, we have had to treat the declination of B, δ_B, being equivalent to that of A, δ_A).

Applying equations (12.5) and (12.6), the distance in declination of A from the centre of the field is:

$$916 \times \cos 65.16 = 384″.80$$

and that of B is:

$$916 \times \cos 53.36 = 546″.66.$$

It therefore follows that according to this preliminary calculation, the difference in declination between the two stars is:

$$384.80 - 546.66 = -161″.86.$$

Since this transit took place north of the field centre, the minus symbol in the answer tells us that B lies north of A. Now, Σ I 57 is a northern hemisphere pair. Hence, in order to obtain B's declination, we need to add 161″.86, or 0°.0450, to that of A:

$$66°.7333 + 0°.0450 = 66°.7783.$$

(If your calculator does not have a facility for automatically converting degrees, minutes and seconds into decimal degrees, simply find the total number of arcseconds and divide by 3600.)

We are now in a position to refine our results by recalculating $\Delta\delta$, substituting the new value for δ_B in equation (12.8). This gives a final figure of 163″.63. We also convert our $\Delta\alpha$ figure into arcseconds, using the mean declination of both stars:

$$13.16 \times 15.0411 \times \cos 66.7558 = 78″.12.$$

After repeating this process for each of the other transits, means are taken of $\Delta\alpha$ and $\Delta\delta$. In this particular case, the results are $\Delta\alpha = 78\rlap{.}''37$ and $\Delta\delta = 164\rlap{.}''85$.

Applying equation (12.10), we obtain the position angle: $\theta = 25\rlap{.}°4$ and from equation (12.11), the separation is $\rho = 182\rlap{.}''5$.

Since $\Delta\alpha$ is positive (B following A) and B lies north of A, we see from Table 12.2 that in this particular case B is in the first quadrant (0–90°), and no further correction to θ is necessary.

According to the WDS, this pair was actually measured by the *Hipparcos* satellite with the following results (1991): $\rho = 182\rlap{.}''4$; $\theta = 25°$. It will be seen that our figures, which are based upon observations made in 1997, are remarkably close. This is certainly a fluke. As a rule, even a large number of transits is unlikely to produce results as seemingly impressive as these. In practice, if you can consistently get within 1° in position angle and 1″ in separation, you will be doing very well indeed. In this particular case, the Zeiss ring micrometer was used on two nights to time six transits across the inner and outer edges of the ring, with the following overall result:

$$\rho = 183\rlap{.}''5; \quad \theta = 25\rlap{.}°2.$$

The position angle result is in full agreement with the *Hipparcos* figure, whereas the separation result differs from *Hipparcos* by less than 1%. This is fairly typical of the level of performance to be expected from the ring method.

For maximum accuracy, a total of not fewer than 10 transits should be taken, preferably spread over several nights. It is good practice to take half the transits near the north edge of the field and the rest near its south edge, taking care not to apply the wrong sign (plus or minus) when calculating $\Delta\delta$. If you have a proper ring micrometer, record the times of appearance and reappearance at its outer and inner edges. In that way, you will be able to refine your results slightly by taking the mean of twice as many timings during each transit. My own experience, as can be seen from the example of Σ I 57, suggests that in this way it should be possible to obtain results to within about 1″ of the true position. Although this is nowhere near good enough for measuring close doubles, it is perfectly acceptable for pairs wider than about 100″.

The rather involved mathematical process of reduction may seem daunting at first sight, but it need not be either laborious or complex if the observer uses a programmable calculator or computer. Once such a device has been programmed to carry out the tedious computations, results can be obtained almost as quickly as the raw timings can be keyed in.

The particular advantages of the ring method are that it requires no special apparatus beyond a stopwatch, needs no form of clock drive or field illumination, can be used with an alt-azimuth telescope as well as an equatorial and is capable of producing consistently accurate results on very wide pairs (separation greater than 100″). It may be worth bearing in mind that although wide and faint doubles lack the glamour of close and fast-moving binaries, they are probably in even greater need of measurement.

The drawbacks of the method, apart from the restriction of its accurate use to very wide pairs, are the rather time-consuming nature of the observations and the elaborate process of reduction. These, although they are greatly reduced by the use of a computer or programmable calculator, can never be entirely eliminated. A Delphi 5 program to carry out this reduction, written by Michael Greaney, is available on the accompanying CD-ROM.

The Chronometric Method

The chronometric method allows a significant increase in accuracy over the ring method. Of comparable antiquity, it requires the addition to the telescope of an external position circle or dial, as well as a single wire or thread mounted at the focus of the optical system. A motor-driven mount is, if not an absolute necessity, at any rate highly desirable. Since position angles are measured directly with the circle, the chronometric method is a hybrid technique rather than a pure transit method. The sole purpose of the timed transits is to obtain differences in right ascension, from which it follows that no calibration exercise is necessary.

An ordinary crosswire eyepiece will serve admirably as the basis of the micrometer. If no such eyepiece is available, a single thread or wire can be mounted in the focal plane of a positive eyepiece, preferably one having a relatively short focal length. The thread must be as fine as possible, ideally no more than 15 microns in diameter. Various materials have been suggested, including nylon or spider's thread. In order to render such materials visible against the dark sky background, some means of illuminating either the field or the thread is essential. A small torch bulb or light-emitting diode may be installed near the objective or inside the eyepiece or Barlow lens. A potentiometer can also be provided so as to enable the observer to vary the level of illumination. Alternatively, at the cost of some degree of precision, the need for a source of illumination may be dispensed with altogether by making the wire relatively thick. I have used a length of 5-amp fuse wire for this purpose. The wire must be stretched diametrically across the field stop and glued in position. The most difficult part of fitting the wire is to keep it under tension so as to ensure that it is perfectly straight. Even then, it is likely to prove rather a crude substitute for an illuminated thread or field.

The position circle or dial can be made from an ordinary 360° protractor, which is fitted to the focusing mount. It must be carefully centred on the eyepiece, to which a pointer or vernier index is attached. The dial must be capable of adjustment by rotation about the optical axis. It is graduated anticlockwise unless the optical system reverses the field, in which case the dial should be graduated in the opposite sense.

Although there is no need to calibrate the micrometer, it is necessary to establish the circle reading that corresponds to north (0°) before measurement begins. One way of achieving this is to find a star near the equator and allow it to drift across the field of view, rotating the eyepiece until the star accurately trails the single thread. Then, leaving the eyepiece undisturbed, adjust the position circle until the pointer indicates a reading of 270° (west). Provided the circle is correctly graduated, it will follow that the zero reading indicates celestial north. By this method, position angles of double stars can be read directly from the PA dial without the need for any correction. However, it is practically impossible to exclude all sources of error in such a home-made device. Quite apart from any defects in the protractor itself, it is unlikely to be perfectly centred on the optical axis. In order to overcome such sources of error, Courtot[1] has recommended the following alternative approach. Adjust the web so that a star drifts along it when the motor is stopped, and note the reading on the dial. Then rotate the eyepiece through 180°, so as to minimise the effects of any centring error, and repeat the process, this time subtracting 180° from the reading. Proceed in this way until you have gathered six readings, and take the mean. The difference between the result and 90° gives the north angle.

Let us illustrate the procedure by reference to Courtot's own example. Suppose that by repeatedly drifting a star along the web we obtain the following circle readings:

East	92°.2	West	273°.3
	92°.5		273°.0
	92°.3		273°.1
Mean	92°.33		273°.13

Subtract 180° from the mean west result: 273.13° – 180° = 93°.13.

Hence the overall mean is (92°.33 + 93°.13)/2 = 92°.73.

Since this corresponds to the true position angle 90°, the north angle is 92°.73 – 90° = 2°.73.

This angle is a correction which will be applied to all subsequent circle readings.

To obtain the position angle of a double star, carefully rotate the eyepiece until the wire is precisely parallel to the pair's axis and note the reading of the PA dial. Then reverse the pointer through 180° and take another measurement. The entire process should be repeated until a total of at least six readings have been obtained. Of these, half will have to be adjusted by 180°. Take the overall mean, remembering to correct for any north angle.

The observer obtains the separation of the pair by timing transits across the wire. At least 20 such timings should be made. There are several variations in the procedure. The simplest way is to set the wire exactly north–south, so that the interval in the times of passage across the wire of the two components corresponds to the difference in RA. The separation is then given by:

$$\rho = \frac{15.0411 \times t \times \cos \delta}{\sin \theta} \tag{12.12}$$

in which t is the mean interval in seconds, δ is the declination of the pair and θ its position angle.

For example, on the night of 2001 August 26, I measured the well-known pair 61 Cyg, with the following results: $\theta = 149°.9$, $t = 1.3384$ seconds.

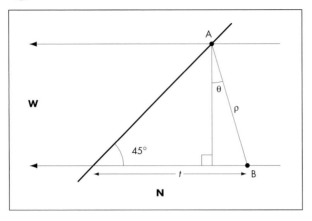

Figure 12.3. With the wire set at 45 degrees to the direction of drift, measure the elapsed time between the transits of stars A and B on the wire.

Since the declination of 61 Cyg is 38°.75, the separation, ρ, is given by (12.12):

$$\frac{(15.0411 \times 1.3384 \times \cos 38.75)}{\sin 149.9} = 31''.3$$

In the case of pairs having a PA close to 0° or 180°, both components will transit the wire more or less simultaneously. There are two ways of overcoming this difficulty. One is to set the web at exactly 45° to the direction of drift (see Figure 12.3), remembering to take into account the north angle. Then, assuming the web is oriented in PA 135°/315° as shown in Figure 12.3, the separation is given by:

$$\rho = \frac{15.0411 \times t \times \cos \delta}{\cos \theta + \sin \theta}. \qquad (12.13)$$

If the web is orientated in PA 45°/225°, the separation is:

$$\rho = \frac{15.0411 \times t \times \cos \delta}{\cos \theta - \sin \theta}. \qquad (12.14)$$

Courtot[1] has suggested an alternative procedure in which the web is placed approximately perpendicular to the pair's axis. The angle, i, between the wire and the direction of drift is read from the circle (making allowance for any north angle). It is positive, increasing from east through south and so on (Figure 12.4). With the telescope clamped a short distance west of the pair, use a stopwatch to measure the time taken for both components to cross the thread. Repeat the process at least ten times, noting the results to two decimal places. Then reverse the wire 180° and take another ten timings.

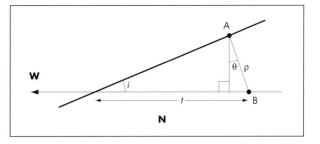

Figure 12.4. Using
Courtot's method with
the wire approximately
perpendicular to the
orientation of the pair.

These timings, together with the declination of the pair and the position angle already determined from the PA dial, enable the observer to deduce the separation, ρ, of the two components:

$$\rho = \frac{15.0411 \times t \times \cos \delta \times \tan i \times \tan \theta}{\sin \theta \times (1 + \tan i \times \tan \theta)}. \qquad (12.15)$$

Using Courtot's own example, suppose that the mean transit interval, t, is 2.386 seconds and the declination of the star is 25°.25. The position angle, θ, has already been measured as 223°.04. Let us further suppose that for the purpose of timing the transits, the web was set with a circle reading of 135°, which corresponds to 45° starting from east. After subtracting the north angle, 2°.73, we find that the web was actually set at $i = 45 - 2.73 = 42°.27$ from east. Then, applying equation (12.15):

$$\rho = \frac{(15.0411 \times 2.386 \times \cos 25.25 \times \tan 42.27 \times \tan 223.04)}{(\sin 223.04 \times (1 + \tan 42.27 \times \tan 223.04))}$$
$$= -21\overset{''}{.}8 \ .$$

The negative value of ρ merely indicates that the companion is west of (preceding) the primary, and the minus sign is therefore ignored.

The main advantage of the chronometric method is that it has greater accuracy than the ring method and can handle pairs down to a separation of as little as 15''. By the careful use of Courtot's variation, this limit may be reduced still further – perhaps even below 10''. Because each transit lasts only a few seconds, it is a relatively quick technique. The reduction procedure, while still somewhat elaborate, is far simpler than the ring method, although the advent of modern electronics has greatly reduced this difficulty in respect of both techniques.

The principal disadvantage of the chronometric method is that for reliable results it demands the use of an equatorial mount. Indeed, it is highly sensitive to misalignment of the mount. If significant errors in position angle are to be avoided, the mount must be accurately set on the celestial pole, with an error of 1′ or less. It follows that the chronometric method is better suited to permanently mounted telescopes than to the portable instruments favoured by many amateurs. Another drawback is that the use of a fine filament necessitates the provision of some form of field or web illumination, which in turn necessarily reduces the working magnitude threshold of the telescope.

Illuminated Reticle Eyepieces

There are now readily available a number of proprietary eyepieces which are supplied by their manufacturers with illuminated reticle systems. They have completely transformed amateur double-star astrometry.[2] The Celestron Micro Guide eyepiece provides a typical example, but other makes are essentially similar (this section refers specifically to the Celestron version). Reticle eyepieces of this type require the use of a motor-driven equatorial mount, with remote slow-motion controls to both axes. This section describes two methods of using the Micro Guide. The first is simple yet very effective, while the more advanced procedure is considerably slower but promises even greater accuracy.

The Celestron Micro Guide is an orthoscopic eyepiece of 12.5-millimetre focal length incorporating a laser-etched reticle and a battery-powered variable illumination system (Figure 12.5). The Meade version uses a different reticle layout (Figure 12.6). In both cases, however, there is a 360° protractor scale at the edge of the field and a linear scale at the centre. The linear scale, which is used to measure separation, is a ruler graduated at 100-micron intervals. Position angles may be determined either by means of an external position circle or, more elegantly and more simply, by using the drift method described in this section.

The first step is to calibrate the linear scale by determining the scale constant, i.e. the number of arcseconds per division. The smaller the constant, the more accurate the measures will be. This dictates as great an

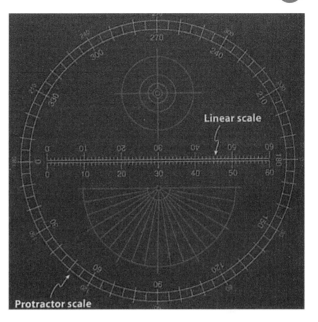

Figure 12.5. The reticle of the Celestron Micro Guide eyepiece. The thickness of the inscribed lines and circles is 15 μm.

effective focal length as possible. Ideally, the focal length should be 5 metres or more, and certainly not less than 3 metres. Since most amateur telescopes have a focal length of between only 1 and 2 metres, it is

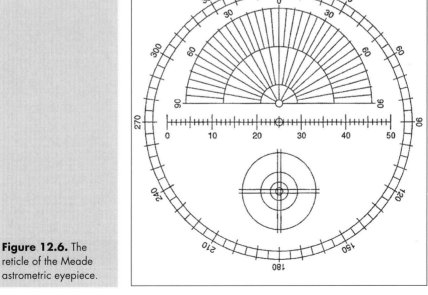

Figure 12.6. The reticle of the Meade astrometric eyepiece.

obvious that a Barlow lens will usually be necessary in order to amplify the image scale at the telescopic focus.

To calibrate the eyepiece, time the passage of a star along the entire length of the linear scale. Select a star that is neither too bright nor too faint – magnitude 5 or 6 will probably be about right for small or medium apertures. In order to minimise the effects of timing errors, choose a star of relatively high declination, but without straying too close to the celestial pole. I have found that a declination of between 60° and 75° is suitable. Rotate the eyepiece until the star drifts exactly parallel to the linear scale. Then use a stopwatch to time the star's journey from one end of the scale to the other. Repeat the process at least 30 times, preferably spread over several nights, and take the mean. To convert the result into arcseconds, multiply by 15.0411 cos δ, where δ is the star's declination. Then divide by the number of divisions in the scale; in the case of the Micro Guide this is 60, but the equivalent scale in the Meade version has 50 divisions. The resulting scale constant, z, will always remain valid for the same optical set-up.

The simpler of the two methods of measuring the separation of a double star is as follows. Rotate the eyepiece until the linear scale is exactly parallel with the pair's axis, ensuring that the primary star is closer to the zero point (or the 90° point in the Meade version) on the 360° protractor scale; although this precaution has no bearing on the separation measure, it will assume importance when it comes to measuring the position angle at a later stage. Then, estimating to the nearest 0.1 division, count the number of divisions separating the two components and multiply the result by the scale constant to obtain the separation in arcseconds.

Measuring the position angle is a slightly more involved process. One way of going about it is to use an external position circle or dial as described in the previous section, but this is actually quite unnecessary.[3] By allowing a star to drift across the field, it is possible to obtain accurate position angles from the 360° protractor scale etched on the reticle itself.

The procedure is as follows: having completed the separation measure, leave the motor running and the orientation of the eyepiece undisturbed so as to preserve the alignment of the reticle. Use the slow-motion controls to bring a star to the exact centre of the field, which on the Micro Guide will be found to

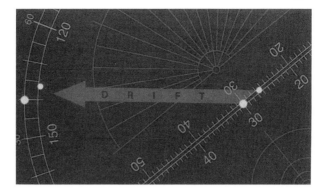

Figure 12.7. Using the simpler method, the pair's separation is measured against the linear scale. The position angle can be found by switching off the telescope's clock drive until the pair drifts to the protractor scale, where the angle is noted. It is important not to bypass or hasten the drift process by using the telescope's RA motor, as unless the polar alignment is perfect, the result will be incorrect.
Reproduced courtesy of Sky Publishing Corporation

lie between the "30" markings on the linear scale. For this purpose, any convenient star will do; it does not even have to be a component of the pair being measured. Once the star is accurately centred, switch off the motor drive and allow the Earth's rotation to carry the star towards the western edge of the field of view. The direction of drift, by definition, corresponds to the true position angle 270°. When the star reaches the 360° protractor scale, switch the motor on and read and record the angle indicated by the star on the protractor scale (Figure 12.7). For a conventional inverted field, the outer (clockwise) set of figures should be used. The inner (anticlockwise) figures are for use with a reversed image, as produced by a right-angle prism. Although the scale is only graduated at intervals of 5°, it is perfectly feasible to estimate to the nearest 0°.5, which is sufficient for all practical purposes.

Subject to one possible correction, the reading indicated by the star shows the position angle of the pair. When using the Celestron Micro Guide, it is necessary to add 90° to the protractor reading in order to arrive at the true position angle. If the final result exceeds 360, just subtract 360 to bring the answer within the range 0–360. With the Meade version, which employs a different layout, no correction is necessary.

As with other techniques of measurement, observations should be repeated over a number of nights and means taken. Used in this way, a reticle eyepiece is capable of making good measures of pairs of any separation lying comfortably within the telescope's resolving ability. It is important to eliminate the effects of parallax by ensuring that the reticle and the star images are focused in exactly the same plane. To achieve this, adjust the telescope focus and the eyepiece dioptre control until you can move your head from side to side without inducing any relative movement between image and reticle.

The beauty of the drift method is that it effectively eliminates index error and places considerably less stringent demands upon the accuracy of the mount's alignment by comparison with a conventional position circle. It follows that this particular technique of measurement lends itself especially well to portable equatorials. Perhaps for that reason, it has become steadily more popular among amateur observers since it was first described in print.[3]

In an alternative, more advanced procedure, the observer uses the reticle eyepiece to measure pairs of angles in each of which both components of the pair are bisected by markings on the linear scale. Employed in this fashion, the eyepiece effectively becomes a degenerate form of filar micrometer. It is a method which produces greater accuracy in the measurement of separation, but it is also slower than the basic procedure already described.

The first step is to rotate the eyepiece until the linear scale is parallel with the axis of the pair to be measured, remembering to ensure that the primary star lies closer to the zero point on the 360° protractor scale. The observer counts the number, n, of whole divisions on the linear scale separating the two components. In the example illustrated in Figure 12.7, it will be seen that $n = 3$. With the motor drive running, the eyepiece is rotated and the slow-motion controls adjusted until a pair of scale markings n divisions apart bisects the two stars as shown in Figure 12.8a. Leaving the orientation of the eyepiece undisturbed, the observer uses the slow-motion controls to bring a star to the exact centre of the field, turns off the drive and notes the angle, θ_1, indicated by the 360° protractor circle at the point where the star drifts across it. In Figure 12.8a, the reading is 60°.

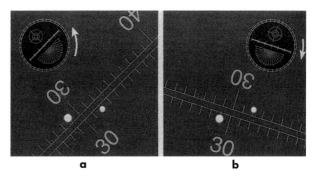

Figure 12.8. The advanced method: **a** measuring θ_1; **b** measuring θ_2. Reproduced courtesy of Sky Publishing Corporation

a   b

Next, the eyepiece is rotated in the opposite direction, past the original position at which the axis and linear scale are parallel, until both components are once more bisected by two markings on the linear scale (see Figure 12.8b). Again, the observer measures the angle, θ_2, as before. In the example shown, the reading is 20°.

If one of the two angles happens to fall within the first quadrant (0–90°) and the other in the fourth quadrant (270–360°), add 360 to the lower of the two figures. This is necessary in order to avoid numerical complications at a later stage in the process of reduction.

The position angle of the pair, θ, is given by the mean of the two angles:

$$\theta = \frac{\theta_1 + \theta_2}{2} \qquad (12.16)$$

to which (in the case of the Celestron version) the 90° correction must be added.

The separation, ρ, is given by:

$$\rho = \frac{n \times z}{\cos \alpha} \qquad (12.17)$$

where n represents the number of whole divisions separating the components, z the scale constant, and α is half the difference between the two angles θ_1 and θ_2:

$$\alpha = \frac{\theta_1 - \theta_2}{2}. \qquad (12.18)$$

In the example shown, $\alpha = 20°$. Assuming a scale constant, z, of 5″, the corresponding separation is therefore

$$\frac{(3 \times 5)}{\cos 20} = 15\overset{''}{.}96.$$

Table 12.3. This observation of $\Sigma 1442$ was made on 2000, Mar 25 with a 21.5-cm Newtonian reflector and Celestron Micro-Guide eyepiece ($z = 6''.25$). Each set of measures occupies a numbered row. The first angle is θ_1, the next a direct PA measure made by the simple "drift" method, and the third θ_2; note all these angles appear in their uncorrected forms. The penultimate column shows the corrected position angle, obtained by adding 90° to the mean of the three preceding entries. The final column gives the separation, derived from θ_1 and θ_2 by the method described in the text. The overall mean position angle and separation appear in the last row

	θ_1			θ	ρ
1	45	68.5	85	156°.17	13''.30
2	49	66	88.5	157°.83	13''.28
3	43	66	94	157°.67	13''.85
4	48	67	89	158°.00	13''.35
				157°.42	13''.45

Again, the procedure should be repeated over a series of nights, and means taken of the position angle and separation. In each set of observations, it is a sensible practice to include a number of direct determinations of the position angle made by the simple method, as shown in Table 12.3.

Because this method of using a reticle eyepiece is insensitive to variations in ($\theta_1 - \theta_2$), it is capable of yielding separation measures far more accurate than those obtained by means of the standard technique. In theory, the precision is not constant, since the uncertainty increases with α. But since it is easier to judge simultaneous bisection at high values of α than at lower values, the competing practical and theoretical considerations probably cancel out.

The range of measurement is restricted by the layout of the reticle. For obvious reasons, the lower limit is set by z, the value of the scale constant. However, it is possible to measure closer binaries by turning the eyepiece through 90° and bisecting the stars with the two long parallel lines, which are only 50 microns apart. Provided the line nearer to the semicircular protractor scale always bisects the primary star, this expedient will also remove any need for a 90° correction; in the case of the Meade version it will, of course, introduce such a correction.

It is the inconveniently short graduation markings on the linear scale that impose an upper limit on the range of continuous measurement. At certain separations beyond about 6z, the observer will find it impossible to bisect both components simultaneously, with the result that gaps begin to appear in the measurement range. For wider pairs, the Barlow lens may always be dispensed with, but this will require the reticle to be recalibrated.

The more advanced method of using an illuminated reticle eyepiece places extreme demands on the observer's patience and dexterity. Not everyone will find the gain in accuracy is really worth the extra time and effort. While it may be useful for occasional measurements, where time is not a consideration, or for the observer who has to make do with a relatively short effective focal length, the amateur who wishes to pursue a systematic programme involving the study of as many pairs as possible will probably prefer to master the simpler technique in conjunction with a telescope having an effective focal length of not less than 5 metres.

Irrespective of the procedure adopted, the illuminated reticle enjoys great advantages over other methods. It is readily obtainable at a reasonable cost and is capable of considerable accuracy.[4] It eliminates index error, is comparatively tolerant of errors in polar alignment and is, therefore, particularly suitable for portable instruments. Its main disadvantage lies in the raising of the magnitude threshold by reason of the illumination system.

Practical Recommendations

Subject to the individual limitations already summarised, any one of the three methods discussed in this chapter is capable of producing results of publishable accuracy. The first two are of particular interest to those who do not wish to buy special equipment. The ring method, although confined to very wide pairs, is ideal for the beginner who wants to attempt measurement without investing in expensive accessories. The chronometric method is more accurate, can handle closer pairs and is perhaps especially suitable for those who enjoy making their own equipment.

For all other purposes, however, the illuminated reticle eyepiece is superior. In the absence of a filar micrometer or equivalent professional apparatus, the observer intending to embark upon a serious programme of visual measurement, with a view to publishing the results, will undoubtedly find the illuminated reticle eyepiece the most practical option.

References

1   Courtot, J.-F., 1999, *The Webb Society Deep-Sky Observer*, **119**, 4.
2   Tanguay, R., February 1999, *Sky and Telescope*, 116.
3   Teague, T., July 2000, *Sky and Telescope*, 112.
4   Harshaw, R., 2002, *The Webb Society Deep-Sky Observer*, **128**, 1

Bibliography

Jones, K.G. and Argyle, R.W., 1986, *Webb Society Deep-Sky Observer's Handbook,* vol. 1: *Double Stars*, 2nd edition.
Martinez, P., 1994, *The Observer's Guide to Astronomy*, trans. Dunlop, S. chapter 13: Double and multiple stars, Cambridge University Press.
Sidgwick, J.B., 1955, *The Amateur Astronomer's Handbook*, section 18: Micrometers, London.

Chapter 13

The Double-Image Micrometer

Andreas Alzner

Introduction

Double-image micrometers generate two images from one incident light source. Instead of setting an external device – like in the filar micrometer – with respect to the object to be measured, the two images with equal brightness are oriented with respect to each other to measure position angle and distance.

The first construction was introduced by P. Muller:[1] he used a birefringent quartz glass crystal. Figure 13.1 shows the orientation of the crystal axes, the incident

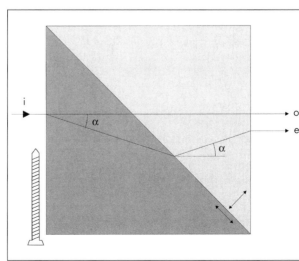

Figure 13.1. The Muller double-image

Figure 13.2. The Lyot–Camichel double image micrometer.

ray i and the ordinary and the extraordinary rays o and respectively. The quartz prism is shifted with the micrometer screw and hence the relative position of the images can be changed.

B. Lyot and H. Camichel[2] suggested a modification with a rotatable calcite plate being the only optical device in the micrometer. Currently this type is the only commercially available double image micrometer made by Méca-Précis, France[3] (Figure 13.2). This uses the spath blade prism (Figure 13.3).

Principle of Operation

the Lyot micrometer the displacement d of the ordinary image with respect to the extraordinary image given in arcseconds is:

$$= \frac{e}{-} \cdot \frac{648\,000 \cdot \sin 2i}{} \cdot \left(\frac{1}{} - \frac{1}{} \right).$$

Figure 13.3. The spath blade prism.

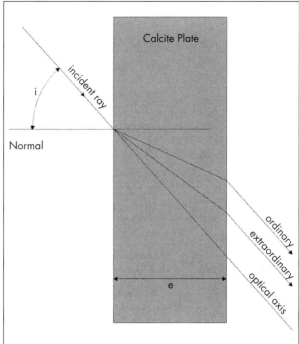

where

e = thickness of the calcite
F = effective focal length of the telescope
i = angle of incident ray
n_e = refractive index of extraordinary ray = 1.48639
(T = 18 °C)
n_o = refractive index of ordinary ray = 1.65836
(T = 18 °C)

The refractive indices vary with temperature and wavelength. The dependency on the temperature T is:

$$\Delta n_e (T) = 1.18 \times 10^{-5} \text{ per } °C$$
$$\Delta n_o (T) = 2.1 \times 10^{-6} \text{ per } °C$$

Measurement

The measurement technique of the double-image micrometer works as follows: The four star images from a double star (primary A, A', secondary B, B') are placed in a configuration the regularity of which can be

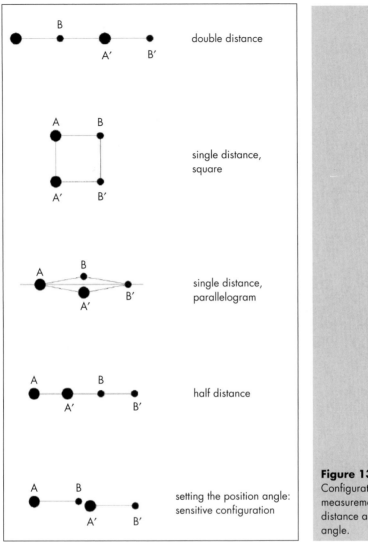

Figure 13.4.
Configurations for measurements of distance and position angle.

For the distance measurements the following arrangements can be used:

double distance (stars on a straight line);

single distance (stars arranged in a square, for close pairs with faint components);

single distance (stars arranged in a parallelogram, if distance is too large for double distance measurement);

For the position angles the calcite is rotated around the optical axis of the telescope in order to line the stars on a straight line; see the sensitive configuration: the closer B and A', the more sensitive.

Coincidences (distance measurement, single distance, stars on a straight line) are to be avoided.

Pros and Cons

Double-image micrometers have some distinct advantages:

- less danger of systematic errors compared with filar micrometers;
- work well with imperfect clock drives;
- for close pairs the distance measurements are easier than with a filar micrometer;
- for close pairs – distance about $2''.0$ and less – with a magnitude difference of not more than 1.5 magnitudes the distance measurements (double-distances, stars on a straight line) are more accurate than with a filar micrometer.

On the other hand the disadvantages are:

- less brightness of the images (loss of about 0.75 magnitudes);
- limitation to relatively close pairs;
- the important parameter e/F (Lyot micrometer) is not easy to achieve.

Accuracy of Double-Image Micrometer Measurements

In order to give a result for the accuracy that can be reached with the Méca-Précis double-image micrometer (DIM) some of the author's measurements 1998–2001 were compared with speckle measurements[4].

The DIM measurements were obtained by using a 32.5-cm f/19 Cassegrain. The limit for clearly resolvable

pairs whilst it is about magnitude 10 for the secondaries for distances more than 1″.2. The limiting magnitudes given above refer to the *Hipparcos* brightnesses given in the *Fourth Catalog of Interferometric Measurements of Visual Binary Stars.*[5] However, the limiting magnitudes are only reached when seeing conditions are good (about 25% of all nights) and the zenith distance does not exceed 40°.

Since orbit grading is a difficult matter and the measurements have been done for many orbits with grade 4 or 5, a subset of the DIM measurements was compared with speckle measurements given in the *Fourth Catalog of Interferometric Measurements.* Single measurements on pairs which could not be separated and measurements which are four years or more off were not considered. Corrections for orbital motion were applied to perform the comparison for the same epoch of observation. Two comparisons are given:

Alzner (DIM measurements) minus speckle measurements on telescopes smaller than 1 m, almost all having been measured with the 66-cm refractor at US Naval Observatory in Washington, DC. The results of this comparison are given in Table 13.1. The distribution of the 119 means as a function of separation used for this comparison is presented in Figure 13.5.

Alzner (DIM measurements) minus speckle measurements on telescopes larger than 1 m (most measurements by Aristide, Docobo, Hartkopf, Horch, Mason and Scardia[4]). The results of this comparison are given in Table 13.2. The distribution of the 63

Table 13.1. Alzner (DIM) – Speckle <1 m comparison, 105 binaries

Parameter	Number or Result
Means DIM	119
Means speckle	119
Observations DIM	290
Observations speckle	307
Observations/mean DIM	2.4
Observations/mean speckle	2.6
Mean of $\Delta\theta$ (degrees)	+0.02
of $\Delta\theta$	±1.07

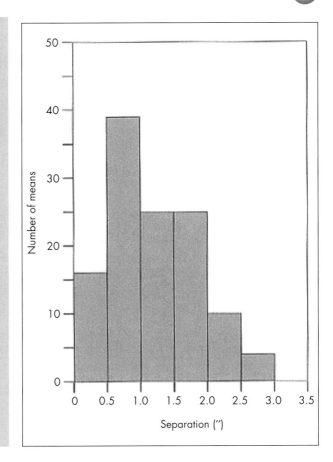

Figure 13.5. The distribution of 119 means used for comparison with speckle < 1 m as a function of separation.

means as a function of separation used for this comparison is presented in Figure 13.6.

Table 13.1 indicates the DIM separations as being about 0″.02 larger than speckle <1 m separations. One

Table 13.2. Alzner (DIM) – Speckle >1 m comparison, 59 binaries

Parameter	Number or Result
Means DIM	63
Means speckle	63
Observations DIM	151
Observations speckle	73
Observations/mean DIM	2.4
Observations/mean speckle	1.2
Mean of $\Delta\theta$ (degrees)	−0.34
σ of $\Delta\theta$	±1.32

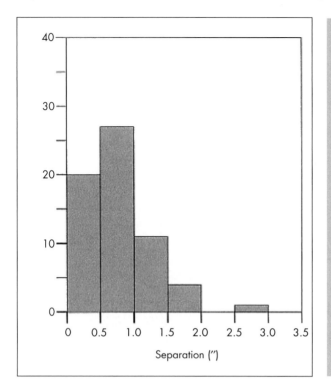

Figure 13.6. The distribution of 63 means used for comparison with speckle > 1 m as a function of separation.

possible reason could be the fact that most DIM measurements were done with magnifications of 620 and 496. These are comparatively low magnifications for separations below 1″.0. There may be a tendency to set the two images of the pair a little further apart in order see the images better. For pairs larger than 1″.0 (62 means, comparison with speckle <1 m) the offset mean (arcseconds) was evaluated as being +0″.005.

Table 13.2 indicates the DIM angles being rotated by −0.34° versus the speckle >1 m angles. However, if the angle for the difficult pair STF 248 is not taken into account, the offset reduces to −0.2°. The comparison of pairs with separations between 0″.40 and 1″.00 (distances with DIM always measured, not estimated!) with speckle observations >1 m (34 means) gives a mean Δρ (arcseconds) of +0″.011.

Prospective users of this micrometer are recommended to read the paper by Christopher Lord[6] which thoroughly discusses the properties and use of the micrometer.

References

1 Muller, P., 1949, *Bull. Astron.* **14**, 177.
2 Camichel, H., 1949, *J. Obs.* **32**, 94.
3 Agati, J.-L. and Huret, R.-G., 1988, *L'Astronomie*, 482–9
4 Alzner, A., Micrometer measurements from 1998.02 to 2002.92, 2003, *Webb Society Double Star Section Circular*, no 11.
5 Hartkopf, W.I., Mason, B.D., Wycoff, G.L. and McAlister, H.A., 2002, *Fourth Catalog of Interferometric Measurements of Binary Stars* (see http://ad.usno.navy.mil/wds/int4.html)
6 Lord, C.J.R., *Measuring Close Double Stars with a Lyot Micrometer*, (http://www.brayebrookobservatory.org) The Méca-Précis Lyot micrometer is available from: Méca-Précis, Zone Industrielle des Sables de Beauregard, 36700, Chatillon-sur-Indre, France. Tel: +33-2-54-02-36-36. Fax: +33-2-54-38-94-54 (e-mail: mecaprecis36@wanadoo.fr).

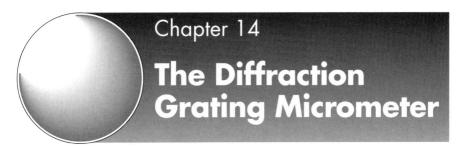

Chapter 14

The Diffraction Grating Micrometer

Andreas Maurer

Introduction

Diffraction influences telescopic images by the effect it has on incoming starlight as we have seen in Chapter 10. It can also be used as the basis for a simple micrometer.

When it comes to measuring the position angles and separations of double stars, sophisticated and expensive precision instruments usually come to mind. However, if you can accept a limited selection of double stars then accurate measurements with very simple devices, the so-called diffraction grating micrometers, are possible. These micrometers, especially in their simplest forms, are very easy and inexpensive to build.

When a telescope object glass or mirror is masked by a coarse grating as shown in Figure 14.1, diffraction of each star image will produce an array of satellite images on both sides of the star in a line perpendicular to the grating slits (Figure 14.5a below). The brighter the star and the wider the grating slits, the greater the number of visible satellites. These satellite images are actually rectangular-shaped spectra but this is only apparent with brighter stars. The central image is the zero order image, the neighbouring satellites are the first order images and so on. For measurement purposes though, only the zero and first-order images of each component are of interest. The basis of this micrometer is that the distance between the zero and first order images is fixed for a given grating and

Figure 14.1. The author's 20-cm Schmidt–Cassegrain equipped with a 50-mm grating. The first-order images are 2.3 arcseconds from the zero-order image. See also Figure 14.5a below. The position angles can be read on a 360° scale.

depends on the separation of the slits. For a given grating therefore, this distance, once determined, can be used to measure both the position angle and separation of double stars.

Experience has shown that gratings whose slit width is equal to the bar width give the best results because this corresponds to the maximum brightness of the first-order images. The critical dimension of a grating is the slit distance, p. The angular separation in seconds of arc between the zero and first order images is given by:

$$z = \frac{206,265 \, \lambda}{(l+d)}$$

where l is the grating slit width (in mm) and d is the bar width (also in mm), so that $p = (l + d)$. The wavelength of the starlight, λ, varies from about 5620 Å (5.62×10^{-4} mm) for an early B star to 5760 Å (5.76×10^{-4} mm) for an early M star but these values depend slightly on the observer, and so λ is known as the effective wavelength. To use the micrometer to its full accuracy each observer needs to determine his or her effective wavelength for a range of spectral types.

The Measurements

As the separation range which can be measured depends on the value of p, to measure all double stars in the range of a given telescope would require quite a number of gratings.

In practice though there is a way to overcome this problem. With a few gratings and some elementary geometry, the basic method can be considerably refined. In this case, a set of four gratings is used with slit distances of 10, 20, 30 and 40 mm. The widths of the bars separating the slits are normally half the slit distance.

The star images and their satellites can be arranged in particular configurations depending on the orientation of the grating. Provided that the pattern is carefully arranged, the grating slit distance and the grating orientation, together with a little trigonometry can deliver quite accurate results for both the PA and separation of the double star being observed. Several star patterns have been proposed by previous observers[1] and the method has been continuously refined. It was extensively and successfully used and described by French and English double star observers in the 1980s.[2,3,4]

Obviously the most convenient method would be a grating with adjustable slit distances, thus minimizing the number of gratings and rendering trigonometric calculations superfluous. Such an instrument had already been proposed by Karl Schwarzschild in 1895.[5] He used three sets of different gratings which he arranged in front of the objective glass of a 10-inch refractor like a roof with rising and descending ridge as shown in Figure 14.2. In this way he could produce variable slit distances, as seen from infinity. The instrument was adjustable by ropes from the eyepiece end.

Lawrence Richardson[6] described a simpler, home-made adjustable interferometer consisting of a flat grating frame which could be tilted in front of a small 4.5-inch refractor. This was the construction which served as a model for the one described here, an easy-to-build, adjustable grating micrometer. It is made of aluminium, board and plywood and is designed for use on the popular 20-cm Schmidt–Cassegrain telescopes. Needless to say the principle of the instrument can also be used on other types of telescopes.

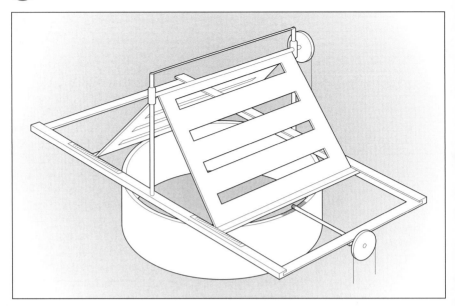

Construction

Figure 14.2. The Schwarzschild adjustable diffraction micrometer used in 1895. Three pairs of interchangeable gratings (p = 70, 40 and 24 mm) were used. Reproduced from Astronomisches Nachrichten by kind permission of Wiley–VCH Verlag

This adjustable micrometer consists basically of two parts:

(1) a rectangular grating frame in front of the telescope objective or corrector lens, which can be tilted with respect to the optical axis and

(2) a flange for mounting this frame with its support onto the objective end of the tube. This flange allows the device at the same time to rotate and its orientation can be read on a 360° dial (Figure 14.3).

The apparent slit distance is varied by inclining the frame which has to be large enough to cover the telescope aperture even when tilted. On the other hand, the frame should not be larger than absolutely necessary in order to keep the instrument size within reasonable limits. Here one has to compromise: as an example, the construction shown in Figure 14.3 works with a 230 × 520 mm frame and the maximum useful tilt is about 65°. The projected slit distance varies as the cosine of the angle of inclination. Therefore the frame-tilt graduation is not in degrees but directly in corresponding cosine values, thus simplifying the reductions.

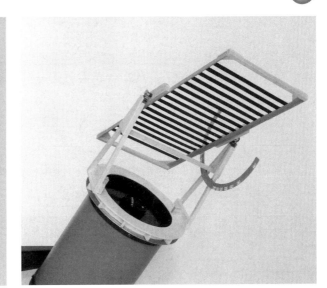

Figure 14.3. A home-made adjustable diffraction micrometer showing the p = 25 mm grating.

For effective diffraction at least three or four slits should be in front of the objective so for a 20-cm tele-scope the largest slits will be about 25 mm wide and arranged 50 mm apart. According to the diffraction formula such a slit distance can thus be used for mea-suring double star separations from 5.5 to about 2.5″. For smaller separations larger telescope apertures are essential. If the 20-cm telescope is to be used for double star separations of up to 10″, say, two grating frames with 50 mm and 25 mm slit distances will do. The smaller grating – used for larger separations – will, when inclined at 65°, produces a projected slit distance of 10.6 mm, which corresponds to about 11″ separa-tion. If wider separations are to be measured a third frame with smaller slit width could be made. However, the stability of the narrow grating strips could become a limiting factor.

In order to get reliable measures, grating frames should be precisely made. The slits and bars should be accurately parallel to each other and also to the tilting axis. Aluminium bars of width 25 mm or alternatively 12.5 mm and 1.0 mm thick are glued onto a frame made of 10 mm aluminium angle and wood. The tilting axis consists of small pivots on each frame side which turn in clamps as shown in Figure 14.4. These clamps allow a frame-exchange within seconds and they also produce just the right friction for the frame to tilt very smoothly.

Figure 14.4. Metal clamps serve as bearings for the grating frames and allow a quick exchange of frames. Note the cosine scale for reading the frame inclination.

Two lightweight side frames support the two grating frame bearings which in turn are fixed to the bottom flange as shown in the photographs. This wooden flange is provided with a cardboard collar on its back, which fits onto the end of the telescope tube. The fit should be tight enough to keep the micrometer properly in place even at low elevations but at the same time not too tight to prevent it being turned around its axis. A collar which is slightly too large is preferable because the desired clearance can then be fixed by inserting some shims of paper or felt. At the collar bottom a 1.5 mm aluminium ring is glued to its rim. This aluminium ring carries a 360° scale or dial and contributes at the same time considerably to the micrometer's stability. This scale, which indicates the double star position angle, is read by a properly set pointer or marking on the telescope tube. Having an outer diameter of 270 mm the scale allows precise reading but if desired a vernier scale could be added. To establish the dial's zero-point the grating slits have to be exactly parallel to the telescope declination axis. In this position the satellite images of a star are aligned north–south. The weight of the micrometer should be kept as low as possible in order not to disturb the balance of the telescope. The instrument shown in the photographs weighs not more than 500 g.

Observing

To make an observation the micrometer is fitted to an equatorially mounted, correctly aligned, and carefully collimated telescope. The 50 mm or the 25 mm grating frame is mounted, depending on the expected separation of the pair to be measured and as high a magnification as possible should be used, preferably 400× or more. The first step is to align the two star components and their satellite images exactly by rotating the whole micrometer (Figure 14.5c). This is called the alignment method and it gives position angles with great precision. Only when the stars and satellites appear properly aligned in a straight line is the position angle read on the 360° dial. At this point it should be noted in which quadrant the fainter star lies in case a correction of 180° needs to be made to the measured position angle. Then the micrometer is rotated exactly 90° further and a configuration as shown in Figure 14.5d will be seen. Now it is time to start tilting the grating frame. This is an easy procedure because when observing with a short 20-cm Schmidt–Cassegrain telescope the grating frame can still be directly reached and operated from the eyepiece end. Great care and judgement is necessary to determine the frame's inclination which produces the correct star configuration. There are two alternative patterns: perfect squares or perfectly right-angled crosses as shown in Figure 14.5e. The idea behind this is, of course, to set the angular distance of the satellite images exactly equal to the double star separation. The mode of operation quickly becomes second nature with the observer and, of course, the larger the series of settings and readings the more reliable the result. In order to compensate for instrument inaccuracies and to increase the precision further, the frame should be swung to both sides and readings on either side on the cosine scale should be made. Furthermore as the satellite images appear on either sides of the stars, two squares or crosses are shown, hence both of them should be judged. As a final verification, the angles A'BA' as well as B'AB' can be checked for perfect orthogonality. Incidentally if a diagonal prism is used the "cross" pattern can be arranged vertically or horizontally for better judgement simply by turning the diagonal. Experience shows that judgement seems to tire quickly so decisions have to be made quickly and alternate glances with either

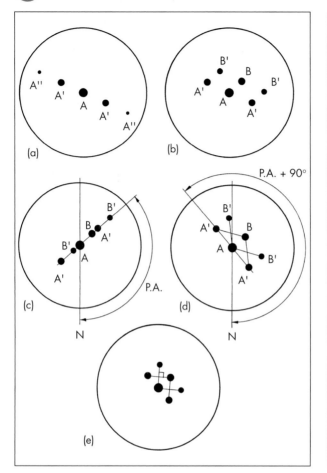

Figure 14.5. Star and satellite patterns as seen in the eyepiece.

eye yield a clearer result instead of staring for too long at the patterns. Only when perfect accord is obtained is the tilt angle cosine read directly from the scale. To get the final value for p the grating's nominal spacing, i.e. 50 or 25 mm, is multiplied by this cosine. Now the diffraction formula can be used to calculate the double star separation, ρ.

It is not necessary to use the "cross" configuration in Figure 14.5E. By swinging the frame, the aligned stars and satellites as shown in Figure 14.5C could for instance be brought directly to exactly equal distances A'–B'–A–B–A'–B' which makes the next step, the instrument's 90° position angle turn, superfluous. Depending on the chosen, lined up star and satellite arrangement, the cosine reading will then need a correction before using the value in the formula. In the described example it has obviously to be multiplied by two. Other alignments – with corresponding correction factors – are possible and thereby the range of the micrometer could be extended considerably. Occasionally, when crowded stars and satellites are lined up in this way, it is perhaps not easy to distinguish stars and satellites. Hence the "cross" configuration as described earlier and shown in Figure 14.5e is preferred, as it works without this added difficulty.

Disadvantages

Diffraction micrometers have one drawback. As the grating consists of bars and slits with the same width, only 50% of the incident light from the double star will reach the telescope optics. Of this, about 50% of the residual light will end up in the zero-order images resulting in a total loss of 1.5 magnitudes compared with the unobstructed telescope. Another 20% goes into each of the first order images, the rest being lost in the additional satellites. Because of these losses the combination of a 20-cm telescope and a diffraction micrometer will allow observations of double stars as faint as about magnitude 7.0–7.5 with components which do not differ too much in brightness.

The diffraction micrometer formula includes the factor λ, the wavelength of light. As the observation is made visually the satellite's exact distance from the primary star depends on the observer's own wavelength sensitivity but also on the stars' colours. The observer's most sensitive wavelength which should be used in the formula has to be established by comparisons with pairs with accurately known separations. A normal figure for λ to start with might be 5650 Å, or 0.000565 mm if p in the formula is in millimetres. This corresponds approximately with the effective wavelength of a white, class A spectral type star.

Accuracy

What about the accuracy of a home-made adjustable diffraction micrometer and what kind of factors will influence a result?

First of all, as with all double star measures, the better the seeing conditions the better the accuracy.

Trying to get results during poor seeing periods will end up in frustration. Good seeing allows high magnifications, which in turn produce large and easy-to-judge star configurations. Then, to obtain accurate results, a series of say 10–12 grating adjustments and readings should be made for a pair; and before the final mean values are determined, such series should even be repeated on consecutive nights. Most crucial for the accuracy of the result is certainly the precise judgement of square and right angle combinations between stars and satellites in the field of view. Equally bright pairs are obviously easier to judge and are thus likely to be more accurate than very unequal pairs.

Also the separation has an influence on accuracy; the closer a pair the higher the magnification needed for a clear interpretation of its satellite arrangement. But the higher the magnification the sooner the seeing can become a limiting factor with its potential negative influence on accuracy. Nevertheless, diffraction micrometer results are surprisingly reliable. Position angles can be obtained with mean errors of 1° and this is good enough to proceed to the next step, the separation measurement. Based on a large number of observations made during acceptable seeing, it can be concluded that for a typical double star the angular separation can be determined typically with a mean error of about ± 2%, but considerably more precise results have often been obtained. Indoor tests under perfect seeing conditions with artificial double stars have shown that still more can be expected from this instrument.

What this means numerically can be shown using two typical examples: for Castor's two bright components (magnitudes 1.9 and 2.9), which in the year 2000 were 3″.8 apart, an accuracy of better than 0″.1 was obtained. In the case of a faint and wide pair, such as STF 1529 in Leo, consisting of components of magnitude 6.6 and 7.4 and separation 9″.5, the separation was determined with an error of less than 0″.2.

Not only are the precision of the construction and careful tuning of star configurations important for the result's reliability, the assumed star wavelengths will also, as the formula predicts, directly influence the accuracy. Catalogues such as the *Bright Star Catalogue* can supply information about the spectral classes of brighter stars, of which Table 14.1 is a small subset. From Richardson's papers, these classes correspond approximately to the following visual wavelengths:

B0: 5620 Å   A0: 5640 Å   F0: 5660 Å
G0: 5680 Å   K0: 5710 Å   M0: 5760 Å

The wavelengths between classes A–F, F–G or G–K do not differ considerably, each step being roughly 0.5%. Hence one might be tempted at first sight to ignore stars' spectral classes altogether, but why ignore useful information when these figures will help to improve the result's accuracy? And here comes a warning: initial diffraction micrometer results with these wavelength figures may perhaps show some strange systematic variations. These can be due to the observer's eye sensitivity or individual interpretation of the star and satellite configurations. Such variations can, as soon as enough experience has been accumulated, be eliminated by personal correction factors.

Is it possible to use the measuring method in reverse to try to calculate and determine the effective observed wavelengths of double stars when their separations and position angles are accurately known from catalogues? With a large database of catalogue data for PA and separations, double star wavelengths can be determined with similar accuracy to separation. Such wavelength

Table 14.1. Pairs with known spectral types near the celestial equator

RA 2000 Dec	Pair	Epoch	PA°	Sep"	V_a	V_b	Sp. Types		Name
01137+0735	STF100	2000	63	23.2	5.21	6.44	A7IV	F7V	zeta Psc
01535+1918	STF180	1999	0	7.7	3.88	3.93	A1	B9V	gamma Ari
03543–0257	STF470	1991	348	6.9	4.46	5.65	G8III	A2V	32 Eri
05350–0600	STF747	1994	224	35.8	4.78	5.67	B0.5V	B1V	
05351+0956	STF738	1997	44	4.3	3.39	5.35	O8	B0.5V	lambda Ori
05353–0523	STF748	1995	96	21.4	4.98	6.71	O7	B0.5V	theta¹ Ori
06090+0230	STF855	1991	114	29.2	5.70	6.93	A3V	A0V	
06238+0436	STF900	1991	29	12.4	4.39	6.72	A5IV	F5V	epsilon Mon
08555–0758	STF1295	2000	4	4.1	6.07	6.32	A2	A7	17 Hya
12413–1301	STF1669	1998	313	5.2	5.17	5.19	F5V	F5V	
13134–1850	SHJ151	1991	33	5.4	6.26	6.76	A0V	A1V	54 Vir
14226–0746	STF1833	1995	174	6.1	6.82	6.84	G0V	G0V	
14234+0827	STF1835	1996	194	6.0	4.86	6.86	A0V	F2V	
14241+1115	STF1838	1997	336	9.4	6.76	6.94	F8V	G1V	
14514+1906	STF1888	2002	316	6.5	4.54	6.81	G8V	K5V	xi Boo
15075+0914	STF1910	1997	212	4.0	6.72	6.95	G2V	G3V	
15387–0847	STF1962	1991	189	11.8	6.45	6.56	F8V	F8V	
18562+0412	STF2417	1993	103	22.6	4.62	4.98	A5V	A5V	theta Ser
19546–0814	STF2594	1991	170	35.6	5.70	6.49	B7Vn	B8V	57 Aql
20299–1835	SHJ324	1991	239	21.9	5.94	6.74	A3Vn	A7V	o Cap
20467+1607	STF2727	2000	266	9.2	4.27	5.15	K1IV	F7V	gamma Del
23460–1841	H II 24	1993	135	6.8	5.28	6.28	A9IV	F2V	107 Aqr

determinations will reveal possible hardware weaknesses, and the overall accuracy can be improved accordingly.

The delicacy of spectral class distinction can also be demonstrated by observing a double star whose components have very different colours. A suitable example is STF 470, consisting of stars of spectral classes G8 and A2 stars and similar brightnesses (magnitudes 4.5 and 5.7). When the images are arranged in the standard "cross" configuration, slightly larger satellite distances for the yellow G8 primary, when compared with the white A2 secondary's satellites, are expected. But even when the two stars, as in this case differ by as much as two spectral classes, it is difficult to detect the slight difference of the first order distances because the two satellite separations still differ by only 1% or so. Hence, for calculating the separation of a double star with components of different spectral class, the mean wavelength of the two stars can safely be used.

Is the diffraction micrometer then even capable of earmarking individual spectral classes? For this purpose, an alternative method, which involves measuring the value of z directly by timing several transits of circumpolar stars can give values of z for a typical grating to an accuracy of about 0.3%. It is necessary to have an eyepiece fitted with a vertical crosswire in order to time the passage of the two first-order images across the centre of the field.[1] Table 14.2 gives a short list of bright circumpolar stars with a range of spectral types which are suitable for this purpose.

Table 14.2. A short list of bright circumpolar stars suitable for determining the value of z

Star	RA2000	Dec2000	V	B–V	Spectrum
HR 285	01 08 44.7	+86 15 25	4.25	1.21	K2II-III
alpha UMi	02 31 48.7	+89 15 51	2.02	0.60	F7:Ib-II
HR 2609	07 40 30.5	+87 01 12	5.07	1.63	M2IIIab
delta UMi	17 32 12.9	+86 35 11	4.36	0.02	A1Vn
HR 8546	22 13 10.6	+86 06 29	5.27	−0.03	B9.5Vn
HR 8748	22 54 24.8	+84 20 46	4.71	1.43	K4III
zeta Oct	08 56 41.1	−85 39 47	5.42	0.31	A8–9IV
iota Oct	12 54 58.6	−85 07 24	5.46	1.02	K0III
delta Oct	14 26 54.9	−83 40 04	4.32	1.31	K2III
chi Oct	18 54 46.9	−87 36 21	5.28	1.28	K3III
sigma Oct	21 08 46.2	−88 57 23	5.47	0.27	F0III
tau Oct	23 28 03.7	−87 28 56	5.49	1.27	K2III

Conclusions

Diffraction micrometers have not only a long and interesting history, they can deliver precise measurements at little cost. If they are made with adjustable slit distances they are easy to use because of easily identifiable star patterns involving a minimum of calculation work. They are therefore ideally suited for amateur observers who want to build a micrometer for their own use.

References

1   Muirden, J., 1980, *Amateur Astronomers Handbook*, 297–304.
2   Pither, C.M., June 1980, *Sky & Telescope*, 519–23.
3   Minois, J., May 1984, *L'Astronomie*, 231–38.
4   Argyle, R.W., 1986, *Webb Society Deep-Sky Observer's Handbook*, (vol. 1: *Double Stars*), 41.
5   Schwarzschild, K., 1896, *Astron. Nachr.*, **139** (3335), 353–60.
6   Richardson, L., *J. Brit. Astron Assoc.*, **35** (1924), 105–6, 155–7; **37** (1927), 247–50, 311–17; **38** (1928), 258–61.

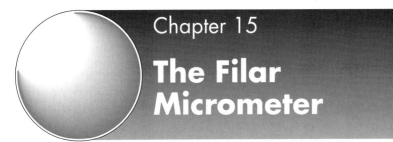

Chapter 15

The Filar Micrometer

Bob Argyle

Introduction

The measurement of double stars is central to the theme of this book and there are many ways of doing this, but this chapter is dedicated to the use of the filar micrometer which has been used seriously since the time of William Herschel. (For a thorough discussion of the history and development of the filar micrometer see the paper by Brooks.[1]) Much of our knowledge of longer period visual binaries depends on micrometric measures over the last 200 years. The filar micrometer is by far the most well-known device for measuring double stars. Its design remains largely the same as the original instrument which was first applied to an astronomical telescope by the Englishman William Gascoigne (c.1620–1644) in the late 1630s. The aim is to use fine threads located in the focal plane of the telescope lens or mirror to measure the relative position of the fainter component of a double star with respect to the brighter, regarding the latter as fixed for this purpose. This is done by the measurement of the angle which the line joining the two stars makes with the N reference in the eyepiece and the angular separation of the fainter star (B) from the brighter (A) in seconds of arc. These quantities are usually known as theta (θ) and rho (ρ) respectively and are defined in Chapter 1.

The basic filar micrometer consists of two parallel wires, one fixed, one driven by a micrometer arrangement, with a third fixed wire at right angles to these two (Figure 15.1). The movable wire must be displaced

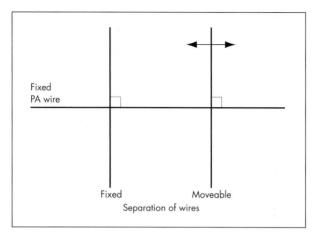

Fixed
PA wire

Fixed   Moveable

Separation of wires

Figure 15.1. The arrangement of wires in a modern filar micrometer.

in the focal plane just far enough from the other two such that it can move freely and yet be in focus. It must also, of necessity, be very thin, preferably smaller than the apparent size of the star disks through the eyepiece. If the focal length of the telescope is too short then a Barlow lens is necessary. This has the advantage of boosting the focal length by two or three times and yet has no effect on the apparent size of the thread. (see Fig. 15.2)

The usual material for the wire is spider thread which was chosen for its fineness and relative ease of availability. (In fact it was a spider making its web in one of his telescopes that gave Gascoigne the idea for the filar.) Replacing spider thread in a micrometer is a relatively skilled job and these days commercially available micrometers use tungsten with a thickness of about 12 microns. The micrometer used by the author has been in regular use for 10 years and the wires have remained correctly set throughout, even though the micrometer has been fitted and removed from the telescope hundreds of times and many thousands of individual settings of the wires made.

In the modern Schmidt–Cassegrain the Barlow lens is a particularly useful accessory. For a 20-cm f/10, for instance, the focal length of 2000 mm is equivalent to a linear scale at the focal plane of 103″ per mm. This means that a 12 micron wire will subtend a diameter of about 1.25″. This is about twice the angular resolution of the telescope so it would limit the user to measuring pairs wider than about 3.0″. Even then the thickness of the threads would make accurate centring of star images difficult.

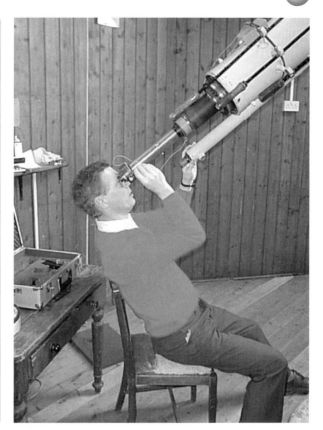

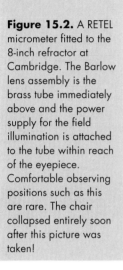

Figure 15.2. A RETEL micrometer fitted to the 8-inch refractor at Cambridge. The Barlow lens assembly is the brass tube immediately above and the power supply for the field illumination is attached to the tube within reach of the eyepiece. Comfortable observing positions such as this are rare. The chair collapsed entirely soon after this picture was taken!

The body of the micrometer must be able to rotate through 360° and its angular position is accurately measured by a circular gauge known as the position angle circle. This is usually graduated in degrees with a vernier available to read to 0.1°.

In the classical brass micrometer, another arrangement called the box screw is usually included. This allows both the fixed and movable parallel wires to be shifted in the focal plane by the same amount. This is useful when the double distance method of measuring separation is employed (described more fully later). For micrometers without this facility (and this tends to include the modern instruments that have become available over the last few years) it is necessary to move the whole telescope to bring the threads into position for double-distance measurement. Alternatively, the method described by Michael Greaney[2] obviates the need to move the whole telescope

After setting the movable wire on the companion and noting the reading, the micrometer is rotated around 180° so that the PA wire bisects the two stars again. The micrometer screw is then turned to move the movable wire across the primary back to the companion. The new reading is then noted and the difference between the two readings gives a measure of the double distance.

As the PA wire bisects the two stars a second PA reading can be taken. Add 180° to this second PA reading if it is less than 180°, or subtract 180° if it is more. The mean of the first and (corrected) second PA readings can be taken as the PA reading for that particular measurement.

Determination of the Screw Constant

This is a rather more difficult task since it is first necessary to determine what the angular equivalent of the linear motion of the micrometer screw is. In the example above we saw that the 20-cm f/10 Schmidt–Cassegrain has a linear scale of 1 mm = 103″ at the principal focus so that if the micrometer has a screw pitch of 0.5 mm per revolution then each rotation of the screw moves the wire 56.5″. It is necessary to subdivide the screw into usually 100 smaller intervals with visual estimates of perhaps one-tenth of each division giving values to 0.001 revolution or 0.06″ in this case. It is necessary to determine this screw value and not to take the manufacturer's data for the focal length of the telescope and Barlow lens. Note however that in those telescopes were the primary mirror is moved to adjust focus then this alters the scale constant and it is therefore important that the scale calibration is checked regularly.

Transits

A commonly used method involves using star transits, but on stars at high declination. With a hand held stopwatch time the transit of a star across the movable wire and note the corresponding value of the micrometer screw. Move the micrometer screw by a fixed amount, say half or one revolution in the direction of the star trail, and time the next transit on the wire. Repeat this for as many revolutions as possible. It will then be possible to calculate a value for one revolution of the screw from all the individual measures. For a star at declination +75 for instance the motion of the star is 15 cos 75″ of time per second so it will take 56.5/15 cos 75 seconds = 14.6 seconds of time to travel the equivalent of one revolution of the micrometer screw in the standard Schmidt–Cassegrain described above. This should be timed to better than 0.5 seconds of time but taking the mean of n revolutions will increase the accuracy of the mean figure by a factor of $\sqrt{n}$. The timings should be repeated on other nights to confirm the figure reached. Further checks at regular intervals are also recommended to see if there is any variation of the screw constant with temperature or with time (due to wear and tear).

Calibration Pairs

Another way of evaluating the screw constant is to measure wide, bright pairs whose position angles and separations are well known and relatively fixed. It will be necessary to have up to a dozen of these pairs spread around the sky so that one can be observed at any time of the year. I use this method, and in Table 15.1 I give a list of pairs with relative positions predicted for 2000.0, 2005.0 and 2010.0. As these pairs change only very slowly the positions for future years can be done by simple interpolation.

Making an Observation with a Filar Micrometer

Position Angle

The measurement of position angle is easiest to make and is usually done first since the measurement of separation depends on the separation wires being perpendicular to the line joining the two stars (Figure 15.1). Position angle is defined as 0° when the companion is due north of the primary, 90° when it is due east and so on. The orientation of the position angle wire can be determined on the sky by several methods; the most common is to set the telescope on an equatorial star, allow the star to drift across the field and rotate the micrometer until the star drifts exactly along the position angle wire. Repeat at the end of the night and the mean of the two values will give the correction to be applied to all readings of position angle made during the night. If for instance at the start of the night the reading is 89°.2 and at the end it is 88°.8 then the mean value of 89°.0 means that +1°.0 needs to be added to each mean position angle taken during the night. Even if the micrometer remains on the telescope it is worth going through this procedure each night.

The measurement of position angle involves setting the PA wire to lie across the centre of the images of each star. It may be difficult to see a faint star under the wire but an alternative of setting the wire tangentially to the two star images is not to be recommended. Another possibility is to use the fixed and movable separation wires set slightly apart, turning them until the line between the stars is parallel to the wires. In this case the exact angle between these wires and the position angle wire needs to be known but once established should remain fixed until the threads need to be replaced.

If using the single position angle wire, it may be necessary instead to turn down the illumination so that the companion can be seen. Several measures of angle should be made depending on the brightness and separation of the pair but it is good practice to move the wire well away from the last determination before making the next measure. This should mean that the readings will be more independent.

It is as well if you are familiar with the position of the cardinal points for the telescope in use. The final position angle, being the mean of each independent setting, may need to be corrected by 180° depending on the quadrant in which the fainter star lies. Remember that in Schmidt–Cassegrain telescopes the cardinal points are a mirror reflection of those in Newtonians and refractors. The use of star diagonals will also add a mirror inversion.

As mentioned above, pairs of accurately known separation and position angle can also be used to calibrate the position angle circle on the sky and a list of some bright ones is given in Table 15.1.

Table 15.1. A list of bright calibration pairs

Pair	RA (2000)	Dec (2000)	Mags	PA 2000	Sep 2000	PA 2005	Sep 2005	PA 2010	Sep 2010
β Tuc	00 31.5	–62 57	4.4, 4.5	168.4	26.98	168.3	26.97	168.2	26.97
ζ Psc	01 13.7	+07 35	5.2, 6.4	62.9	22.81	62.8	22.77	62.8	22.73
θ Pic	05 24.8	–52 19	6.3, 6.9	287.7	38.14	287.7	38.14	287.6	38.14
δ Ori	05 32.0	–00 18	2.2, 6.8	0.2	52.45	0.1	52.45	0.1	52.45
γ Vel	08 09.5	–47 20	1.8, 4.3	220.4	41.22	220.4	41.22	220.4	41.21
ι Cnc	08 46.7	+28 46	4.0, 6.6	307.4	30.39	307.4	30.38	307.5	30.37
Σ1627	12 18.2	–03 57	6.6, 7.1	195.6	20.00	195.6	20.00	195.6	19.99
24 CBe	12 35.1	+18 23	5.0, 6.6	270.2	20.18	270.2	20.18	270.2	20.18
α CVn	12 56.0	+38 19	2.9, 5.6	228.5	19.34	228.5	19.32	228.5	19.30
ζ UMa	13 23.9	+54 56	2.2, 3.9	152.3	14.44	152.5	14.45	152.6	14.45
κ Lup	15 11.9	–48 44	3.9, 5.7	143.1	26.45	143.1	26.42	143.1	26.40
ν Dra	17 32.2	+55 11	4.9, 4.9	311.0	62.07	311.0	62.08	310.9	62.09
θ Ser	18 56.2	+04 12	4.6, 5.0	103.6	22.38	103.6	22.40	103.6	22.42
16 Cyg	19 41.8	+50 32	6.0, 6.3	133.3	39.56	133.3	39.62	133.2	39.69
o Cap	20 29.9	–18 35	5.9, 6.7	238.4	21.86	238.4	21.86	238.4	21.85
β PsA	22 31.5	–32 21	4.3, 7.1	172.2	30.37	172.2	30.38	172.2	30.39

β Tuc Both stars are close pairs in a large telescope.
ζ Psc The companion is a close pair in a large telescope
θ Mus The primary is a close pair in a large telescope.
δ Ori The primary is a close pair in a large telescope.

Separation

The most common technique for the measurement of distance is called the "double-distance" method (see Figure 15.3). Basically the fixed wire of the micrometer is placed on the primary star and the movable wire on the companion. The reading of the movable wire is noted. The telescope and micrometer screw are then moved until the fixed wire is placed on the companion and the movable wire placed on the primary star. The difference between the two positions of the screw is twice the separation of the pair in millimetres (or whatever unit the screw is calibrated in). This is repeated several times, depending on the difficulty of the pair. The separation of the pair in arcseconds is then calculated by $k(r_2 - r_1)/2$ where k is the screw constant and r_1 and r_2 are the mean values of each separation setting. I make four double distance measures for wide pairs and up to six measures for close pairs. This procedure, like that of the determination of position angle, is repeated for several nights before a mean value is determined for each. It is better to make the measures of separation close to the position

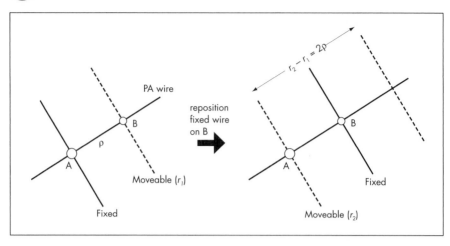

angle wire, since if the separation wires are not strictly parallel then the measure of separation will be in error and in any case the images will be better near the centre of the field.

Figure 15.3. Double-distance method of determining separation.

An alternative method by Michael Greaney[2] is illustrated in Figure 15.4. The CD-ROM contains Delphi 5 programs for calibrating and using filar micrometers.

Illumination

The best way of illuminating the field of the micrometer is to direct a low but variable light onto the wires, i.e. bright wire illumination. In some micrometers, notably the RETEL which uses a red LED, the field is bright and the wires are seen in shadow. Whilst red is usually regarded as the colour least likely to reduce the

Figure 15.4. Alternative double distance measurement (Greaney).

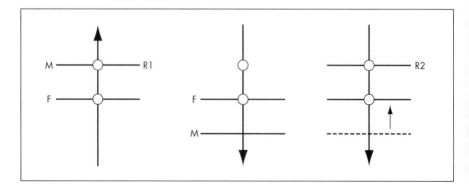

effectiveness of the eye and distract the observer, some astronomers prefer a different arrangement. Paul Couteau uses a white light to illuminate the wires whilst Wulff Heintz prefers yellow. Observers will appreciate that white light is definitely not recommended for field illumination.

Calibration Pairs

Table 15.1 lists bright and wide pairs whose position angles and separations can be predicted with sufficient accuracy to calibrate a filar micrometer. The data used for this has been taken from the *Observations Catalogue* at USNO, courtesy of Dr B. D. Mason. The *Catalogue* contains all published observations irrespective of accuracy so some of the measures have been excluded from consideration. Sixteen bright pairs have been chosen to cover the north, the equator and the south for all times of year. The southern pairs are considerably less frequently observed and the predicted positions are therefore less reliable.

Although some of these pairs are real, if very slow-moving, binaries the observed arc is less than 5° in most cases and so motion is assumed to be linear. A weighted, least-squares straight line fit to the data has been made in all cases with the weighting being made arbitrarily. It was decided to give micrometer measures a weight equal to the number of nights whilst photographic measures (and also *Hipparcos* and *Tycho* measures where applicable) were given a weight of 50. As an example Figure 15.5 shows the observations of ζ UMa (= Σ1744) from around 1820 to the present day, more than 350 in total. The effect of long sets of photographic measures made after 1950 is to dominate the fit but the earlier measures also fit the line reasonably well lending confidence to the predicted positions. In separation, there has been no significant change since observations began.

In each case in Table 15.1 it was first necessary to correct the observed angles for precession, bringing them up to the year 2000.0. The values given in the table for future years have also been corrected for precession to those epochs allowing an immediate comparison to be made with observations.

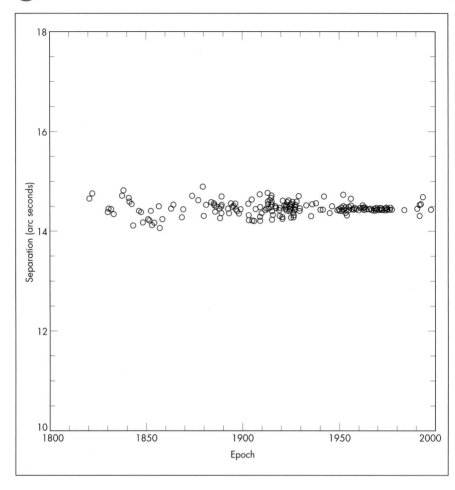

Errors of Measurement

Figure 15.5. The measures of Mizar in separation and position angle 1820–2000.

When any quantity is measured errors can arise in the process. These can be two kinds. Firstly, random or accidental errors which are caused by natural fluctuation when making, for instance, a number of measurements of the separation of a double star with a filar micrometer. If you take say six readings at each position of the movable wire, the numbers will differ slightly. Taking the arithmetic mean of these numbers yields a figure which can be taken to be a fair representation of what the value being measured should be. This can be converted into an angular separation in the usual way. If the pair being measured is a binary star of known separation then if the same measurement is

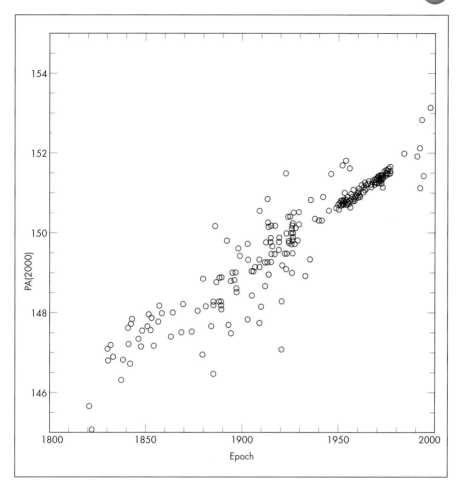

Figure 15.5.
(continued)

repeated on several other nights and the subsequent mean values all indicate a greater separation than expected then you may suspect the existence of a systematic error. It may be that the current orbit is not predicting the correct separation for the time of observation but it could also mean that the screw value for the micrometer is not correct. If the screw value is based on a single standard pair then there is room for systematic error to come in – it may be that the separation assumed is not correct. This can be tested by observing other standard pairs to see if the same screw value is obtained. If it is then the binary orbit can be suspected.

This is a particularly interesting and vital area which needs to be considered regularly if micrometric measures are to be regarded as stable and reliable.

Sources of Error

Positioning of Wires on a Star

The two graphs below illustrate the comparison of micrometer measures which I made (observed measures) with accurate measures of the same stars made with speckle interferometers and by the *Hipparcos* satellite and referred to below as the reference measures. When making these comparisons it is vital that the epochs of measurement agree as closely as possible, otherwise the comparisons are not valid due to orbital motion (or proper motion) during the interval.

Figure 15.6a shows the differences between the observed and reference separations. In this case the sense is (observed-reference) so that for the closest pairs (below about 1″ or so) the measured separations are too large. This is not an uncommon feature of measurement by micrometer and it is particularly useful for anyone doing orbital analysis. Whilst the raw measures are published as they stand, in the case of a particularly careful orbit calculation, it pays to try and assess the "personal" error of the observers and then to apply correction to the observed positions. In practice this tends not to happen much because suitable refer-

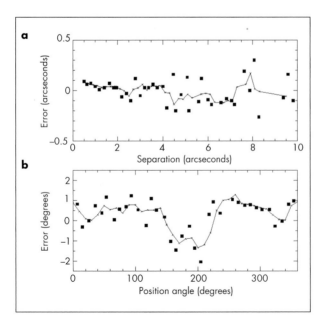

Figure 15.6. a The error of a mean separation with separation. The solid line represents a running average. **b** The error of a mean position angle with position angle. The solid line represents a running average.

ence measurements have not been available for comparison. This has changed for the better recently with the publication of the speckle results from the USNO (see the reference in Chapter 25) where suitably accurate and up-to-date measurements are available to enable observers to check their personal equation.

There is a large scatter at larger separations and this is due to a combination of the paucity of standards at these separations and fewer measures which I have made. Of the points in Figure 15.6a some 210 pairs below 2″ are compared, dropping to 69 pairs between 2 and 4″ and only 31 pairs between 4 and 10″.

What can be seen from the graph in Figure 15.6a is a tendency for me to measure the closest pairs (0.5–2″) as rather wider (about 10%) than they really are and from about 2″ and wider there is not much systematic error to be seen.

In Figure 15.6b the graph shows the situation for the observed position angles for the same pairs as Figure 15.6a. Here there is clearly an anomaly at about PA 180°. This is where the two stars appear nearly vertical in the eyepiece. Although it is recommended to place the eyes either parallel to or at right angles to the line between the stars it is more uncomfortable to do the former so I conclude that using the eyes parallel to the line between the stars results in an error in position angle of about −0.5 to −1° when the stars are within about 40° of the vertical. Another way to avoid this is to use a prism in conjunction with the eyepiece to allow the field to be rotated by 180°. By making another set of position angle measures here the mean value should then be free of this particular bias.

Accuracy of Reference Pairs

When using reference pairs to calibrate micrometers it is better not to use binaries because it is much easier to obtain accurate relative coordinates from wide pairs. In most cases these stars have been measured by *Hipparcos* or *Tycho* and there are plenty of measures going back over time which indicate any significant binary motion.

Errors in the Micrometer Screw

Each measure I make involves at least eight settings of the micrometer screw – typically 3000 settings per year.

It is reasonable to suppose that some wear and tear or backlash might make itself noticeable at some stage so regular checking should be made. This can be done by plotting the scale values derived from standard stars with time.

Availability of Filar Micrometers

For many years filar micrometers had been unobtainable and although the occasional classical brass example does appear they tend to get snapped up by collectors and placed on the shelf. Over the last 10 years, however, a number of firms and individuals in the UK and USA have produced commercial instruments and there are at least two sources of supply at the time of writing. (See the references for further details.)

The RETEL micrometer is made in the UK from duralumin alloy and consists of a fixed and movable parallel wires and a PA wire at 90°. The movable wire is driven by an engineering micrometer capable of about 12 mm of travel and readable to 0.001 mm using the vernier. The PA circle is calibrated in 1° intervals and again a vernier allows this to be improved to 0.1°. The wires are made from artificial fibre and are 12 μm thick which means that for short focus telescopes a Barlow lens is needed to reduce the apparent size of the wires in the eyepiece. Spider thread is better in terms of thickness but it is difficult to fit and needs regular replacement. The man-made fibre is extremely durable – I have had no breakages in 13 years of continual use involving many thousands of individual settings.

The van Slyke micrometer is made in the USA from a solid block of aluminium and again features an engineering micrometer to drive the movable wire whilst a range of optional extras such as digital readout are also advertised. Unfortunately, as this was being written the micrometer has been transferred to the manufacturer's discontinued catalogue but was still available as a custom order.

A comparison between the two made by Andreas Alzner can be found on the Webb Society web page.[3]

References

1 Brooks, R.C., 1991, *Journal for the History of Astronomy*, **22**, 127.
2 Greaney, M.P., 1993, *Webb Society Quarterly Journal*, **94**, 22
3 Alzner, A. (http://www.webbsociety.freeserve.co.uk/notes/dsretel.html)
   The RETEL micrometer is available from Retel Electro-Mechanical Design Limited, 37 Banbury Road, Nuffield Industrial Estate, Poole, Dorset, BH17 0GA, UK. Contact Mr L. Reynolds – Tel: (01202) 685883; Fax: (01202) 684648.
   The van Slyke Engineering Filar micrometer is still available on special order – see http://www.observatory.org/turret.htm

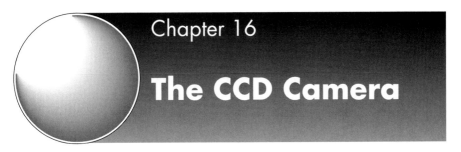

Chapter 16

The CCD Camera

Doug West

Introduction

The charge-couple device (CCD) camera has opened up many new possibilities for the amateur and professional astronomer.[1, 2] The study of double stars has also benefited from the advent of the CCD camera. The measurement of position angle, separation, and magnitude can all be derived from a single short (few seconds) exposure with the CCD camera. With a typical camera available to the amateur and a 20-cm telescope (see Fig 16.1), pairs as faint as $V = 16$ can easily be measured. The increased sensitivity of the camera to light allows pairs much fainter to be measured than would be possible visually or photographically.

Like most innovations, there are some drawbacks and limitations inherent with this technology. Visual double star pair measurements taken with a micrometer are complete at the time of observation, but this is not the case with a CCD. After the image is taken with a CCD, then the processing of the images begins. In the simplest case, reasonable measurements of three quantities, position angle, separation, and magnitudes can be obtained from images that have no post observation processing. If your measurements require the most accuracy available from the image then more elaborate processing is required. The most accurate measurements require the observer to be knowledgeable, though not an expert, in techniques of astrometry and photometry.

The purpose of this chapter is to give the observer guidance in using a CCD camera to take double star measurements. As it is with most things, those individuals that are diligent in learning and experimenting with the telescope and camera can reap double star measurements that are both rewarding to the observer and of scientific value.

CCD Camera Basics

The CCD camera allows multiple astronomical objects to be recorded on the same image and the detector is linear so that photometric information can be obtained. Unlike photograph film, which has grains positioned in a random fashion on the film and a variety of grain sizes, the CCD pixels are equally spaced in rows and columns. This consistency of spacing allows the position of stars to be determined very accurately. Once the image is taken with a CCD camera the image is read by the computer and converted to an electronic file. The raw image in electronic format can now be processed in a variety of ways to enhance the image, for example, the images can be stacked, rotated, filtered, re-formatted, have quantitative measurements taken, etc.

A commonly used CCD camera is the Santa Barbara Instruments Group (SBIG) ST-8E (shown in Figure 16.2). The heart of the SBIG ST-8E CCD camera is the KAF-1062E CCD chip. The KAF-1062E has a detection surface that is composed of an array of 1530×1020 pixels, which are each 9 microns square. The CCD chip is thermoelectrically cooled to 25 °C below the ambient air temperature to suppress electronic noise generated within the chip. Each pixel converts the energy in the light into electrons that are trapped in the pixel. After the exposure is complete, the camera's electronics reads the number of electrons in each pixel. This information is then converted to a black and white image on the computer screen.

Drift Method of Measurement

Once the CCD image has been taken and the normal post processing completed, such as, removing the dark

Figure 16.1. The author and his 8-inch Schmidt–Cassegrain

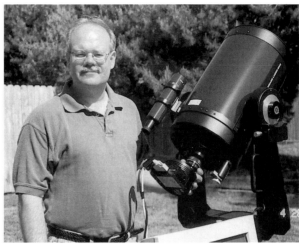

frame and dividing by the flat field, then the position angle and the separation of the pair can be determined. The simplest method to determine the position angle and separation is the drift method. This method consists of taking an image of a pair and during the exposure turning the telescope's clock drive off. This produces an image like the one in Figure 16.3. The lines produced by the drifting star images define the east–west line in the image. A protractor can be used to measure directly from the image the position angle.

Figure 16.2. The SBIG ST-8E CCD.

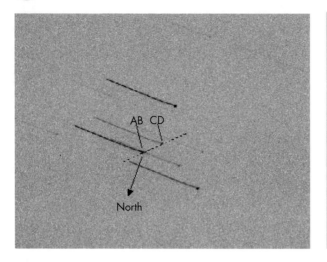

AB CD

North

Figure 16.3. A drift field image of WDS 02473+1717.

The measurement of the separation requires that the distance between two points on the image be calibrated in arcseconds per unit length. The calibration can be accomplished using the standard pairs from Table 15.1 in Chapter 15.

To illustrate this method, consider the quadruple system WDS 02473+1717 (Figure 16.3). The AB and CD pairs are too close for the CCD image to resolve, therefore, they are treated as single stars in this example. The WDS lists the AB pair at V magnitude 8.57 and the CD pair at magnitude 9.71. The AB–CD separation is 104″.59 and the position angle is 135°.0. The image in Figure 16.3 is a 30 second exposure taken with the author's 8-inch Schmidt–Cassegrain telescope with a ST-8 CCD camera.

The simplest approach to measuring the pair consists of first enlarging the image as much as practical and printing it. With a straight edge, draw a line joining the AB and CD points. Use a protractor and measure the angle between the east–west line (formed by the drift of the star's images) and the AB–CD line. This turns out to be 45°. Since the east–west line is perpendicular to the north–south line, 90° can be added and the result is 135° – amazingly good agreement with the WDS! For widely separated pairs this method works quite well. Closely spaced pairs and pairs where the stars nearly align with the east–west line cause problems for this method.

Once the position angle has been measured, the next step is the separation. Measurement of the separation requires that the field be calibrated in arcseconds per pixel. This can be accomplished using the standard pairs from Table 15.1 or by using the method covered in the next section of this chapter. The separation is determined by counting the number of pixels between the two stars. This requires the x and y coordinates of the two stars be known. In this case, the position of the AB pair is at (521, 547) and the CD pair is at (582, 518). The distance between the pair is found using the following formula:

$$\text{Separation (pixels)} = \sqrt{(521-582)^2 + (547-518)^2}.$$

The resulting separation is 67.5 pixels. The telescope/CCD combination has 1.52″ per pixel. This gives a separation of 102.7″. This is within 2% of the WDS value of 104.59″.

The drift method of measurement is the probably the simplest way to measure double stars. However, this simplicity has a price, that is, the accuracy of the measurements is not as good as the CCD camera is capable of delivering. To access the full potential of the CCD camera image the method in the next section is required.

Astrometry Method of Measurement

The position angle and separation are calculated from the measured right ascension and declination positions of the star pair.[3] The following equations define the relationship between the pair separation (ρ) and the position angle (θ):

$$\rho = \sqrt{(\alpha_b - \alpha_a)\cos^2\delta_a + (\delta_b - \delta_a)^2} \qquad (16.1)$$

$$\theta = \tan^{-1}\left(\frac{\delta_b - \delta_a}{(\alpha_b - \alpha_a)\cos\delta_a}\right) \qquad (16.2)$$

where α_a is the right ascension of the brighter star and α_b is the right ascension of the dimmer star in the pair, δ_a is the declination of the brighter star and δ_b is the declination of the fainter star.

To illustrate this method, an example is in order. Figure 16.4 is a negative CCD image of the field of the quadruple system WDS 07131+1433. This is a routine system with only three of the four stars visible in the figure. The separation of the B and D star is below the

resolution of the CCD frame and they appear as one star. Table 16.1 gives the measurements from the WDS *Catalog*.

The first step in the process is to determine the right ascension and declination of at least three stars in the field. These stars are used to define the positions of the remainder of the stars in the field of view. A variety of software packages are readily available to perform the required astrometry calculations. Some examples of software that performs astrometric measurements are: Axiom Research's MIRA, Herbert Raab's Astrometrica, John Rogers' CCD Astrometry, Project Pluto's Charon, Bob Denny's Pinpoint, and BdW Publishing's Canopus. These software packages typically use the *Guide Star Catalog*, *Tycho*, and the USAO 1.0 catalogs.

The author used the MIRA software and the positions from the *Tycho* catalog, which are labelled 1–4 in Figure 16.4, to determine the positions of the stars labelled A, B, and C in the same figure. The positions of

Figure 16.4.
Negative CCD image of the field of the quadruple system WDS 07131+1433.

Table 16.1. WDS data for WDS 07131+1433

Pair	Date	θ	ρ	Magnitudes
AB	1979	353	63.7	9.1, 11.1
BC	1920	311	10.0	11.1, 11.9
BD	1905	332	3.5	11.1, 11.0

Table 16.2. Position of reference stars

Star	Tycho	RA(2000)	Dec(2000)
1	774 345 1	07h 12m 54s.64	+14 24′ 33″.2
2	774 1053 1	07h 12m 48s.97	+14 36′ 03″.5
3	774 535 1	07h 13m 20s.19	+14 48′ 07″.7
4	774 821 1	07h 12m 19s.88	+14 40′ 36″.3

Table 16.3. Measured positions for WDS 07131+1433

Star	RA(2000)	Dec(2000)
A	07h 13m 05s.05	+14° 32′ 57″.85
BD	07h 13m 05s.12	+14° 34′ 01″.42
C	07h 13m 04s.46	+14° 34′ 09″.10

the reference stars are listed in Table 16.2. It is best if catalog positions for any of the stars in a double star system are not used as a reference star, the reason being that the measurement process tends to introduce systematic errors in positions. Calculating the position angle and separation effectively removes the effects of the systematic error.

The closest pairs that can be measured with certainty using the 8-inch are about 3″ (providing the difference in magnitude is 2 or less). This is limited by the pixel size and the seeing.

Using the four reference stars in Table 16.2, the three visible stars in WDS 07131+1422 are measured. Table 16.3 gives the positions of the stars in the quadruple system.

With the positions known for the three stars in the system and equations (16.1) and (16.2), the position angle and separation are now calculated. Comparing Tables 16.1 and 16.4 you see a difference in the separations and position angles. These differences are due to the errors in both previous and current (Epoch 2000.787) measurements and the actual position changes of the stars due to their proper motions.

Table 16.4. Measured PA and separations

Pair	(2000)	(2000)
AB	0°.93	63″.58
BC	308°.69	12″.28

Photometry

One of the big advantages of the CCD camera is that it allows the measurement of a star's magnitude very accurately. Knowing the brightness (or magnitude) of a star is a fundamental property that is very important to astronomers and double star work. Precision of 0.01 magnitude can readily be obtained with a CCD camera.

The simplest type of photometry[1,4] is differential photometry. In differential photometry, a star with unknown magnitude is compared to a star with known magnitude, and from the difference in flux the magnitude of the unknown star can be computed. Using the following equation the magnitude of an unknown star can be measured:

$$\Delta M = M_1 - M_2 = 2.5 \ \log \ (f_1/f_2)$$

where f_1 and f_2 are the flux of the two different stars and M_1 and M_2 are the magnitudes that correspond to the fluxes, respectively.

As an example, consider the stars labelled A and 4 in Figure 16.4. Star #4 is HIP 34810 which has a V band magnitude of 9.10. We wish to measure the magnitude of star A. To do this we start with a CCD frame that has had the dark frame removed and has been divided by a flat field image. The owner's manual for the CCD camera should cover the topics of dark frames and flat fields.

The next step is to measure the fluxes of the A and #4 stars. The simplest approach is to use the region of interest (ROI) capability of the CCD software or a secondary image processing software package. The ROI function in the software typically provides the mean and standard deviation of the region specified by the cursor. The ROI should encompass the entire star with little additional sky coverage – this will ensure a reasonably accurate flux measurement. The mean value of the star should be recorded and the mean value of a region near the star should be measured. This second measurement is of the background sky. The flux of the star is the difference between the two measurements:

Flux(star) = (mean counts in star's ROI) –
(mean counts in background ROI).

For stars A and #4:
Flux(star A) = 102.98 – 78.73 = 24.25 counts
Flux(star 4) = 132.2 – 78.04 = 54.16 counts

$$\Delta M = M_1 - M_2 = -2.5 \ \log \ (24.25/54.16) = 0.87.$$

Using $\Delta M = 0.87$, and $M_A = M_4 + \Delta M$, the magnitude of the A star becomes: $M_A = 9.10 + 0.87 = 9.97$. From the *Tycho* catalog, star A is TYC 774 41 1 which is magnitude 10.01. The error is 0.04 magnitude from the *Tycho* catalog. This is a typical error to expect for this method of differential photometry.

The photometric accuracy can be improved by taking multiple differential measurements and then averaging. Different reference stars can also be used – this will average out the effects of errors in reference star magnitudes and any variability in the reference stars. The ROI method of differential photometry is one of the simplest methods and more advanced methods can be used to obtain more accurate measurements.

References

1 Buil, C., 1991, *CCD Astronomy: Construction and Use of an Astronomical CCD Camera,* Willmann-Bell, Inc.
2 Ratledge, D. (ed.), 1986, *The Art and Science of CCD Astronomy*, Springer-Verlag.
3 Kitchin, C., 1998, *Astrophysical Techniques*, 3rd edn, Institute of Physics Publishing.
4 Budding, E., 1993, *An Introduction to Astronomical Photometry*, Cambridge University Press.

Further Reading

Elliott, G.A., 1998, *The Double Star Observer*, **4**, (2), 5–13.
Henden, A. and Kaitchuck, R., 1982, *Astronomical Photometry*, Van Nostrand Reinhold.
Martinez, P. and Klotz, A., 1998, *A Practical Guide to CCD Astronomy*, Cambridge University Press.

Photometry and Astrometry Software

CCDSoft/TheSky Software. http://www.bisque.com
MPO Canopus. http://www.minorplanetobserver.com/htms/mpocanopus.htm
Project Pluto's Charon. http://www.projectpluto.com/charon.htm

MIRA. http://www.axres.com/
Astrometrica. http://www.astrometrica.at/
John Rogers CCD Astrometry http://www.Camarillo
Observatory.com

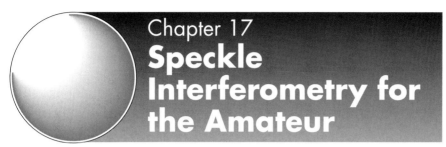

Chapter 17
Speckle Interferometry for the Amateur

Nils Turner

Introduction

There is a bit of a revolution going on in the amateur astronomy field. In the pages of a typical amateur astronomy publication one finds advertisements for telescopes with apertures greater than 50 cm, high-quality charge-coupled device (CCD) camera systems, image processing software packages, and even intensified CCD cameras! In addition, the articles in these publications assume a more knowledgeable reader, with more references to technical concepts such as the Fourier transform's role in forming an image. With the ubiquity of the fast personal computer and the above mentioned equipment, the time is ripe for them to try speckle interferometry.

Binary Star Astrometry

Binary star astrometry is all about two quantities, angular separation and position angle. In a binary star system, the angular separation is the angle between the two stars subtended on the sky, while the position angle is the orientation of the axis connecting the two stars with respect to north. By convention, north is at 0° and east is at 90°.

Making measurements from a perfect image is easy. Due to the circular aperture of the aberration-free telescope that generated this perfect image, an unresolved star along the optical axis has a point-spread function (the two-dimensional intensity distribution in the image of energy from the star) that is radially symmetric, with a bright central core, and a succession of concentric rings. This pattern is known as an Airy disk (see Chapter 10).

The Airy disk is radially symmetric: there is a point that can be determined to be the centre. If there are now two Airy disks, one from each unresolved star in a binary star system, getting the astrometric information is straightforward – determine the angle and distance of separation in the units of the image, and use knowledge of the scale and orientation of the image with respect to the sky to get astrometric information suitable for publishing.

This scenario changes with the introduction of seeing – short time-scale variations of the image intensity introduced by the atmosphere. If the separation of the two stars is smaller than the size of the seeing disk, the astronomer risks being unable to distinguish between the two stars.

Speckle Interferometry

The speckle interferometry technique involves processing the image before the seeing disk becomes the dominant feature of the image. This is usually done by taking short exposures (typically 15 milliseconds or less) of the star and analysing the pre-seeing-disk structure for information. Under high magnification, the area around the star reveals itself to be many "specks" (or "speckles") moving at random within a relatively confined area. Figure 17.1 shows a speckle pattern from a 4-m telescope.

Looking at many of these short exposure images in succession, one sees a sort of "shimmering" effect. This is due to interference between the individual speckles, hence the term, "speckle interferometry".

Labeyrie[2] laid much of the groundwork for using speckle interferometry in a scientific capacity. Although it has been known since the time of Fizeau and Michelson (late nineteenth, early twentieth century) that resolution information lost due to the atmosphere can be regained through the use of interferometric techniques, it has only been available on the brightest of objects.

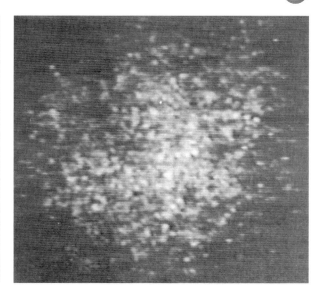

Figure 17.1. A single speckle frame of the star κ UMa on the KPNO 4-m. Adapted from McAlister et. al.[1] Printed by kind permission of the American Astronomical Society

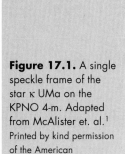

Several groups dabbled in binary star speckle interferometry throughout the mid-1970s, but with the commonplace usage of image intensifiers in observational astronomy by the late 1970s, the true potential of speckle interferometry in binary star science was revealed to the astronomical community.[3]

Turbulence

The underlying cause of the speckles seen in high magnification, short exposure images is atmospheric turbulence. Coulman[4] wrote an excellent distillation of atmospheric turbulence and how it applies to astronomy. To simplify for the binary star astronomer, atmospheric turbulence can be thought of as many pockets of air of subtly different temperature moving across the column of air defined by the telescope aperture, and in the direction of the object. In a time averaged sense, most seeing conditions (an observer's working measurement of turbulence) can be described by two numbers: r_0, a measurement of the diameter of the typical pocket of air passing in front of the telescope aperture; and τ_0, a measurement of how long a typical pocket of air "influences" the wavefront getting into the telescope. The value of r_0 has a wavelength dependence:

$$r_0 \propto \lambda^{6/5} \qquad (17.1)$$

Typical values of r_0 (at 550 nm) and τ_0 are 10 cm and 15 milliseconds, respectively.

In practical terms, one might hear of the seeing being "1.6 arcseconds". This is actually a measurement of the "seeing disk", the diameter at the full-width half maximum (FWHM) of a Gaussian (equations of the form $\exp(-x^2)$ representation of the stellar intensity. To translate this into a value of r_0, ten Brummelaar[5] calculates:

$$r_0 = 1.009 \, D \left(\frac{\lambda}{\theta_{seeing} D} \right)^{6/5} \qquad (17.2)$$

where λ is the wavelength, D the objective diameter of the optical system, and θ_{seeing} the FWHM diameter (in radians) of the seeing disk. So, 1.6″ seeing, as viewed at 550 nm through a 50 cm telescope, translates to r_0 of about 5 cm.

Measuring r_0 requires determinations of interferometric visibilities at different exposure times and extrapolating back to the exposure time of 0. Determining object visibilities is a bit beyond the scope of this chapter, so suffice it to say that for the speckle astronomer, r_0 determination is a more qualitative exercise. When the highly magnified image of the object meanders "slowly" along its random path, the conditions are said to be those of "slow seeing". If the object appears more animated in its travels, conditions are described as "fast seeing". How r_0 affects the design and operation of an astrometric speckle system is described by Mason.[6]

Speckle Interferometry in Practice

So, how does one use images of swirling speckles to determine binary star astrometric data? To answer that question, it helps to know a little bit about the Fourier transform and function convolution, and how they play out in the imaging process.

Imaging Process in Brief

Figure 17.2. The principle behind speckle imaging and the autocorrelation procedure. Courtesy of Dr. H. A. McAlister. Reproduced by permission from Sky Publishing Corporation.

If one assumes a point-source object, the corresponding image intensity, after going through the atmosphere and the optical system, is the convolution of the object source intensity with the Fourier transform of the atmospherically disturbed telescope pupil (the intensity distribution seen in "collimation" mode – most often seen by the amateur astronomer aligning a Newtonian telescope). The convolution process alters the original, ideal object source by a combination of atmospheric distortions and telescope aberrations.

The physical manifestation of this procedure can be seen from Figure 17.2. The top right image shows a short exposure, highly magnified star image taken through a narrowband filter. The envelope of individual speckle images is the seeing disk, in this case about 1″ in diameter. Inside the seeing disk can be seen the individual speckles but more importantly pairs of

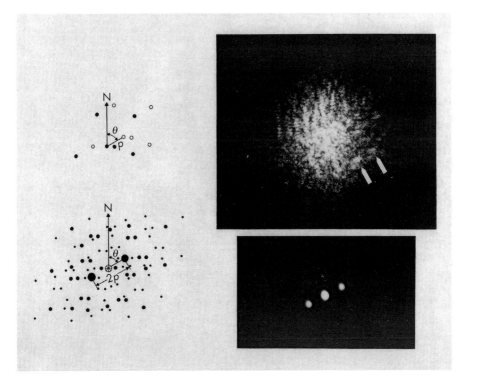

speckles which have the same orientation and separation (such as that indicated by the arrows) can be made out. Each pair of speckles represents the components of the binary star imaged at *full* resolution of the telescope (in this case about 0.03″. The separation of the pair of stars is about 0.27″ and the position angle 293°. The position angle and separation could be measured directly from this frame but the power of the speckle method is that it uses many pairs of speckles to increase the reliability of the measurement.

The top left image shows five typical pairs of speckles. These move randomly inside the seeing disk but the relative separation of the speckles in each pair always remains the same. The essence of the Fourier transform is that it assesses the frequency of spatial separations of the speckles – each speckle from every other speckle. It can be seen from Figure 17.2 that there is a wide range of separations between speckles of different pairs but underlying this the most often occurring separation is that between the speckles representing the binary. However, because we are considering the separation of each speckle from every other then there are just as many pairs of speckles at PA 113° as there are at 293° so there is an ambiguity. Resolving this ambiguity is treated later in the chapter.

Calculation of the Fourier transform of the above-described image intensity produces a picture frequently referred to as the power spectrum of the image. The power spectrum represents the distribution of image "power" among the available "spatial frequencies". As an example, consider a stream of (one-dimensional) audio data. If this data represents two pure musical tones, the combined audio waveform will be the mixture of two sinusoids with different periods and (most likely) different amplitudes. If one were to take the Fourier transform of a finite section of this mixed waveform, the resulting diagram would reveal the constituent waveforms (and their relative amplitudes). Returning to the two-dimensional realm of astronomical imaging, the frequencies encountered are spatial rather than temporal. In the case of imaging a binary star system, the most likely spatial frequency to occur in the individual speckle snapshots is the separation between the two stars. In this case, the combined power spectrum of many snapshots will show bands of light and dark. The crest-to-crest distance is mapped to the separation of the two stars, while the axis perpendicular to the bands represents the position angle.

The reader interested in learning more about the Fourier transform and convolution is strongly encouraged to seek out the book by Bracewell.[7] Alternately, most undergraduate level textbooks on optics will have sections describing these two concepts as well – see, for example, Klein and Furtak.[8]

Making the Measurements

To aid further the determination of binary star astrometry, the Fourier transform can be used yet again. By calculating the transform of the power spectrum, these bands of light and dark can be converted into a sequence of three co-linear, circularly symmetric peaks. Binary star astronomers call this picture the autocorrelogram. The autocorrelogram consists of a large central peak and two smaller peaks, one on either side of the central peak, exactly 180° apart. Figure 17.3 shows a typical autocorrelogram.

These peaks are the result of a random process, which gives them a Gaussian profile. Therefore, centres are easy to determine. The distance between the centres of the central peak and one of the other peaks

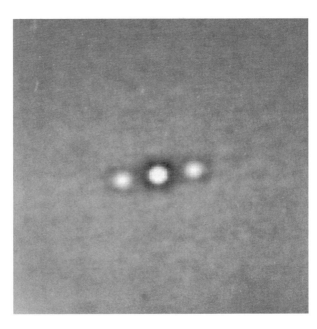

Figure 17.3. A background subtracted autocorrelogram of the binary star κ UMa. Adapted from McAlister et al.[1] Printed by kind permission of the American Astronomical Society

gives the separation angle. The angular orientation of the line of three peaks gives the position angle, though with a 180° ambiguity.

This process of creating the power spectrum and the autocorrelogram can be done using analogue or digital techniques. Gezari et al.[9] describe a typical analogue autocorrelogram generating process. They start with a standard 35 mm film camera attached to an image intensifier tube. The camera is equipped with a rapid film advance system to more efficiently use telescope time, though, in the strictest sense, it is optional. The camera records a sequence of short exposure, high magnification images. This film is then developed and negative reversed (so the film is now a positive image). Each individual frame is stepped through a laser-illuminated optical system (employing the classical aperture/image relationship of coherent optics) and individually exposed onto a separate emulsion to form the power spectrum. This process is done again to form the autocorrelogram. This is usually a slightly toxic process in that an index-matching fluid to the film has to be used to keep laser scattering from micro-scratches from ruining the power spectrum and auto-correlogram emulsions. For traditional 35 mm film, this fluid is usually a variant of the standard dry cleaning fluid, naphthalene.

This whole process can be done digitally if digitised frames of the individual speckle frames are available. While one can digitise the individual photographic frames, video frames from a bare CCD or intensified CCD are the more likely source. Taking the Fourier transform of each individual frame, co-adding them all to form the power spectrum, and taking the Fourier transform again produces the autocorrelogram.

In the digital realm, there is a shortcut, a way to go straight from individual speckle frames to the autocor-relogram. This is done by correlating every pixel in the individual speckle frame with every other. To see this, imagine an "autocorrelogram canvas" four times the size of an individual speckle frame. For each individual frame, a value of one is assigned to pixel values above a certain threshhold, and zero below. For each above-threshhold value in the individual frame, its pixel address is aligned with the centre of the canvas and the frame is added to the canvas (remember that the frame is now a collection of ones and zeros). Morphologically, this procedure produces a diagram much like that of generating the autocorrelogram through the use of

Fourier transforms. In fact, taking into account the multi-bit nature of the data, and adjusting the threshhold value, these numerical techniques can be shown to be the same.

Directed Vector Autocorrelogram

Now, what does one do about the 180° ambiguity in the typical autocorrelogram? Bagnuolo et al.[10] describe a variation of the direct image-to-autocorrelogram numerical technique that resolves the quadrant ambiguity for most binary star systems. Instead of merely assigning the above-threshhold value to one, its multi-bit value is retained. The concept of aligning the frame multiple times on a larger canvas is replaced by an analysis of each of the unique pairs within the frame. In the case of the former concept, the net effect is to sample each unique pair twice, aligning the frame on each component of the pair. It is very easy to see that this will lead to a peak on either side of the central peak, exactly 180° apart. In the case of the latter concept, each unique pair is sampled only once, aligning the pair only on the brighter value. This will tend to emphasize the outlying peak that, along with the central peak, will define the true position angle of the binary star system. However, if both stars are about the same brightness, the method breaks down. This variation of the autocorrelogram is called the directed-vector autocorrelogram. Figure 17.4 shows a surface plot of a directed-vector autocorrelogram.

Speckle Sensitivity

As was alluded to above, r_0 strongly influences the sensitivity of any speckle observation. The "faster" the seeing is, the shorter the camera exposure time has to be to get enough non-overlapping speckles to process. A shorter exposure time implies a brighter magnitude limit. Also, to first order, a larger collecting aperture does not increase sensitivity. A larger telescope creates a physically smaller Airy disk as well as more speckles. In order to be able to use the techniques of speckle

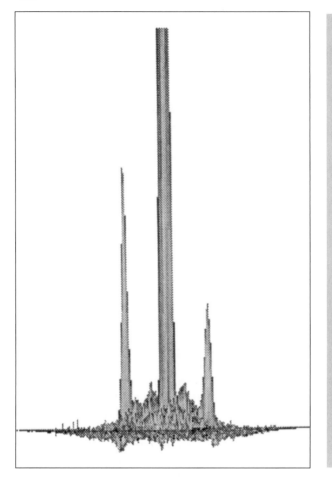

Figure 17.4. A background subtracted directed-vector autocorrelogram of the binary star 21 Oph. The central peak has been truncated for clarity. Adapted from Bagnuolo et al. Printed by kind permission of the American Astronomical Society

interferometry, one has to magnify the image and/or shorten the exposure time to get enough non-overlapping speckles, thereby offsetting the increased light-gathering power of the larger aperture size. A larger aperture merely gives better resolution. In 20 years of using speckle interferometry for binary star research, Hartkopf[11] finds that the best results occur when the number of detector pixels across the central peak of the Airy disk is between 10 and 30. For wider systems (those where the speckles of the individual components barely overlap, and wider), better sensitivity can be gained by pushing the number of pixels below 10. However, it is not a good idea to go below 2 – one runs the risk of losing a significant amount of light to the space between the pixels. There is nothing special about these values, and the curious user is encouraged to experiment with them. They were determined empirically – autocorrelograms of the same collection of binary star systems were taken over a wide range of image scales and the results compared. Chapter 10 describes how to determine the size of the central peak of the Airy disk, and using that size to design a system will be described in the earlier section on Eyepiece Projection.

Equipment Considerations

Telescopes

Speckle interferometry works with any telescope design. Because the necessary field-of-view is quite small, no complex optical system is necessary to get good astrometric results. The most important descriptive number of a telescope system is the objective focal length, f_{obj}. If it is not listed on the telescope, an approximate value is easy to calculate. It is the focal ratio multiplied by the objective diameter. In the case where the main telescope is only one part of a more complex optical system, f_{obj} can be calculated from a simple measurement in the image plane of an object of known angular extent. For example, the angular extent of a certain object is known to be 2.3″. After passing through the optical system, the size in the image plane is measured to be 14.5 μm. So,

$$f_{obj} = \frac{14.5\ \mu m}{2.3\ \text{arc sec}} \times \frac{206,265\ \text{arc sec}}{1\ \text{rad}} \times \frac{1\ mm}{1000\ \mu m}$$
$$= 1300.4\ mm$$

Note that in the above calculation, the radian "units" have been ignored. Radians are a unitless measure. In order to do optical calculations, all angular measurements need to be converted to the "natural" units, radians.

CCD Cameras

Which camera or camera system to choose is a matter of cost, convenience, and/or complexity. While one could go for a standard, large-format, digital imaging system, the long readout times of the CCD array make processing thousands of frames very inconvenient. In addition, many of the imaging CCDs are not equipped with accurate or reliable enough shutters. High frame rate digital CCDs (Dalsa makes this type of camera, see the list of suppliers at the end of the chapter) are another option. The disadvantage is in complexity. These cameras typically require building a custom electronic interface and writing a low-level device driver to get the image data into a location in computer memory. So, this leaves video cameras. For speckle interferometry they fall into two classes, intensified and non-intensified. Both types are useful; they simply need to have a standard video output (RS-170 or PAL). Which to choose is mostly a matter of price – a non-intensified video camera is in the range of US$250–300, and an intensified video camera is in the range of US$2000–15,000.

When choosing a video camera for use in a speckle system, be sure to get information on the physical size of the CCD detector, not just the number of pixels along each dimension (though this is handy to know as well). This video camera will be used in conjunction with a frame grabber (described below) which will rescale the output video. If this information is not available, all is not lost. Simply make sure the speckle system design has enough flexibility to optimise the Airy disk size to the output pixel size.

Any video camera will do, even a colour camcorder. It just needs to output RS-170 (NTSC if it is colour) or PAL analogue video. For sensitivity purposes, a monochrome video camera is better. For the adventurous (and moderately wealthy), a video camera with adjustable gating can increase performance on brighter objects or under faster seeing conditions. An RS-170 video camera outputs 30 frames per second, a PAL camera, 25. Normally the CCD detector exposes for the whole frame time. Through either creative "charge dumping" (reading out the CCD several times during the frame time, and only "remembering" the last readout) or an LCD shutter, the time the detector sees the sky can be less than an individual frame time while keeping the video frame rate unchanged. This process is known as gating.

The dedicated amateur with a bit of money to spend might consider an intensified video camera. Intensification adds an additional 4–5 magnitudes of sensitivity, and thereby increase the number of available objects 10- or even 100-fold. There is a penalty.

Intensifiers are electronically noisy. They use phosphors to create electrons from the original photons, amplify these electrons about a million-fold through a series of high voltage cathodes, and use phosphors again to create many photons out of the amplified electrons. An input of a photon creates an output of about a million photons. If, during the early stages of amplification, a stray thermal electron enters the series of cathodes, it gets amplified right along with photon generated electrons. A detector peering at the output of the intensifier cannot tell an object photon from a thermal electron. For binary star astrometry, using an intensifier limits the magnitude difference between the stars to about 3.[6]

The final issue about the camera is its size and weight. Lightweight and compact cameras are easier to use on all types of telescopes. In addition, many cameras come with the camera head (containing the detector) and the controlling electronics as separate units connected by a cable. This makes for a system that is more modular and easier to handle.

Eyepiece Projection

Frequently, with the typical amateur telescope, the most appropriate method to match the Airy disk size with the detector pixel size is the technique known as eyepiece projection. An eyepiece is placed between the chain of telescope optics and the detector which steepens the effective input angle to the detector, increasing the effective focal length of the optical system. Mathematically,

$$f_{\text{effective}} = \frac{d_{\text{eye}} \times f_{\text{obj}}}{f_{\text{eye}}} \qquad (17.3)$$

where d_{eye} is the eyepiece–image plane distance, and f_{eye} is the eyepiece focal length.

Using the example optical system from the earlier section on Telescopes, in conjunction with a 28 mm focal length eyepiece, a 2000 mm effective focal length can be attained by setting d_{eye} to 43.1 mm:

$$d_{\text{eye}} = \frac{f_{\text{effective}} \times f_{\text{eye}}}{f_{\text{obj}}} \qquad (17.4)$$

$$= \frac{2000 \text{ mm} \times 28 \text{ mm}}{1300.4 \text{ mm}} = 43.1 \text{ mm}.$$

Eyepiece projection systems can be bought commercially (see the list of suppliers at the end of the chapter) or custom-built using standard machine shop tooling.

In practice, d_{eye} is fixed to give the desired $f_{effective}$, and the whole projection apparatus is moved to adjust the focus. Finally, keep in mind that not all eyepieces will physically fit in a commercial system. The wide angle, large eye relief eyepieces (TeleVue Nagler, Meade Ultra Wide Angle, etc.) have significantly larger housings than the more traditional Kellners, Orthoscopics, and Plössls.

Filters

For pure astrometry, the choice of filter is somewhat arbitrary. Roughly, the number of speckles is proportional to the diameter of the telescope aperture and inversely proportional to the wavelength (assuming a not too broad bandwidth):

$$n \propto \frac{D}{\lambda}. \qquad (17.5)$$

The speckle size has a wavelength dependence as well as a weaker bandwidth dependence. A speckle is essentially an image of the Airy disk, dominated by the central peak. Equation (10.1) shows the functional form, which is for an infinitesimally narrow bandwidth. If a finite bandwidth is used, each wavelength "bin" creates its own Airy disk, each of a slightly different size. The net effect is to "fuzz out" the Airy disk edges, which makes the speckles bigger, and merges more of them together. In addition, for wide bandwidths, the atmosphere will chromatically split the object light, creating an elongated speckle. This effect is more prominent with larger telescopes and objects at a significant angle down from zenith. A filter with a narrower bandwidth lessens this problem and makes astrometry more precise.

Astronomers mostly use the Johnson UBV[11] and Cousins RI[12] filter systems. These are fairly wide bandwidth filters. Another system that is frequently used is the Strömgren $ubvy$[13] – see also Crawford and Barnes[14]. These are quite a bit narrower and may give more precise astrometry. Additional information about astronomical filter systems can be found in the *General Catalogue of Photometric Data*[15] (visit http://obswww.

unige.ch/gcpd/gcpd.html). Again, for pure astrometric work, any filter that produces good, sharp speckles will suffice.

Frame Grabbers

The final piece of equipment to discuss is the frame grabber. The frame grabber digitises each RS-170 or PAL frame and saves it to a location in computer memory, where the DVA program applies its thresholds and extracts the relevant data. With the memory and speeds of today's computers and frame grabbers, the DVA program can process nearly every frame in real time. Because RS-170 and PAL signals are a sequence of analogue scan lines, the frame grabber will effectively resize the pixels. For example, the video camera may have 780 detector pixels across the CCD. This row of pixels gets converted to an analogue signal, the scan line. The frame grabber then re-digitises the scan line into a row of, say, 256 pixels. Recall from the section on sensitivity that the sensitivity of the system is based on the number of detector pixels across the Airy disk. The frame grabber pixelation value chosen is simply a matter of processing speed.

The choice of frame grabber is a question of price and custom programming. At present, work is underway to port a DOS DVA program using an Imaging Technology Plus frame grabber (no longer available) to a Linux DVA program using the Matrox Meteor frame grabber (visit http://www.chara.gsu.edu/~nils/dva.html). The Matrox Meteor costs about US$500, and a clone is available from Omnimedia Technology Inc. for about US$400. The Matrox Meteor is chosen because a device driver has been written for the Linux operating system. Matrox also makes a frame grabber called the Meteor II. The Meteor II is not yet supported under Linux.

An Example System

As an example, it is decided to build a speckle system around a typical Meade or Celestron 8 inch, f/10 Schmidt–Cassegrain telescope. For these systems, $f_{obj} =$

2000 mm. This will be used in conjunction with a CCD video camera that has 540 pixels across a scan line and 15 ìm pixels. In addition, an eyepiece where $f_{eye} = 13$ mm is available for use. First, calculate the necessary $f_{effective}$ to get the central core of the Airy disk to span about the right number of pixels. To get the core to span 10 pixels, the Airy disk core needs to be about 150 ì in size. From equation (10.1),

$$f_{effective} = \frac{D_{Airy}D}{(2.44)\lambda} = \frac{150\ \mu m\ 203.2\ mm}{(2.44)\ 550\ nm} = 22712\ mm.$$

Using equation (17.4),

$$d_{eye} = \frac{22712\ mm \times 13\ mm}{2000\ mm} = 148\ mm.$$

To calculate the field-of-view (fov) in seconds of arc along the scan line,

$$fov = \frac{(Number\ of\ pixels) \times (pixel\ size)}{f_{effective}}$$
$$= (540\ pix) \times \frac{15\ \mu m}{1\ pixel} \times \frac{1\ rad}{22712\ mm} \times \frac{206225\ arc\ sec}{1\ rad}$$
$$= 74\ arc\ sec.$$

As can be seen from equation (17.4), to shorten d_{eye} for a given $f_{effective}$, either use a shorter focal length eyepiece or a longer focal length telescope. One can use a Barlow lens to get a longer effective focal length. In certain cases, changing to a longer effective focal length or shorter focal length eyepiece is not practical. In these cases, one may have to replace the eyepiece in the projection setup with a microscope objective. Because microscope objective focal length parameters are not described like those of an eyepiece, using one typically involves a bit of trial and error. Start with the lowest power (typically ×5) objective. If the image scale is still not quite right, change to a higher power objective. A bit of simple machining will be required to make an adaptor to fit the objective into the projection apparatus. Remember that the threads of the objective are pointed towards the camera.

Figure 17.5 shows a photograph of the prototype amateur speckle camera, built and tested in early 1992. It uses a microscope objective and a Philips monochrome CCD video camera. A description of its performance can be found in Turner et al.[16] (visit http://www.chara.gsu.edu/~nils/1992cadm.conf..577T.pdf).

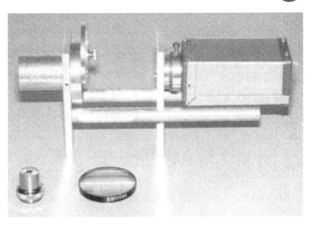

Figure 17.5. The amateur speckle camera built in early 1992. Photo courtesy of Sky Publishing Corporation.

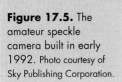

Scientific Programme

Astronomers at the United States Naval Observatory (USNO) maintain a list of all the visible measurements, published in the literature, of binary stars. They make available on the world wide web a summary list (http://ad.usno.navy.mil/ad/wds/wds.html. Also at the USNO web address are lists of single stars, and multiple star systems with orbital elements. By observing single stars, one can test new data reduction techniques. Systems with known orbital elements are useful for determining the exact scale and orientation of the optical system on the sky.

The above mentioned summary list makes a good starting point for a project of study. For the data to be useful, it is necessary to calibrate the data accurately. The best way to do this is to look at known, slow moving systems – a system that has not changed position angle and separation significant for the last 200 years is a good candidate for scale and orientation calibration. It is necessary to observe several of these systems before taking apart or modifying the optical system.

Appendix

Equipment

This is a partial listing of equipment available to build a speckle system. It is by no means complete. Because this author lives in the United States, the list has a distinctly North American slant. The interested astronomer is encouraged to seek out local resources when building a system. Amateur magazines are good resources for the local distributors of the products of international manufacturers (such as Celestron or Meade) as well as local manufacturers of similar products.

Eyepiece Projection Systems

Commercial systems are available from Meade, Celestron, and Apogee. Contact a local distributor. The Large Scale Vendors section below lists the contact information for these companies.

Eyepieces

Eyepieces for eyepiece projection systems are available from numerous manufacturers and distributors. Again, be sure the eyepiece purchased will fit the projection system. Some manufacturers include the usual suspects, Meade, Celestron, Tele Vue, and Apogee. Again, the Large Scale Vendor section lists the contact information for these companies. In addition to the well-known international companies, the following companies also market eyepieces.

Orion Telescopes & Binoculars, PO Box 1815-S, Santa Cruz CA 95061 USA. (http://www.telescope.com)

Pentax, 35 Inverness Dr East, Englewood, CO 80112, USA (http://www.pentax.com)

University Optics, PO Box 1205, Ann Arbor, MI 48106, USA (http://www.universityoptics.com)

Filters

Filters can be purchased from the companies listed below. Several of these companies (Andover, Barr Associates, Custom Scientific, CVI Laser, and OCLI) are speciality houses, dealing only in filters, specializing in custom applications. As a result, they are more expensive. The best bets for inexpensive (but non-standard) filters are Edmund Scientific (contact information is listed in Large Scale Vendors below) and Lumicon. The contact information is listed below.

Andover Corp., 4 Commercial Dr, Salem NH 03079, USA (http://www.andcorp.com)

Barr Associates, Inc., 2 Lyberty Way, Westford, MA 01886, USA (http://www.barrassociates.com)

Custom Scientific, Inc., 3852 North 15th Ave, Phoenix, AZ 85015, USA. (http://www.customscientific.com)

CVI Laser Corp., 200 Dorado Pl SE, Albuquerque, NM 87123, USA. (http://www.cvilaser.com)

ISI Systems, 3463 State St PMB283, Santa Barbara CA 93105, USA. (www.imagingsystems.com)

Lumicon, 6242 Preston Ave, Livermore CA 94550 USA (http://www.lumicon.com) – forced to suspend operations

Optical Coating Laboratory, Inc. (OCLI), 2789 Northpoint Parkway, Santa Rosa CA 95407, USA. (http://www.ocli.com)

Oriel Instruments, 150 Long Beach Blvd, Stratford CT 06615 USA (http://www.oriel.com)

High-Speed Digital Cameras

This is a source for a high frame rate (upwards of 1000 frames per second) digital camera. Use of this camera would require major modification of the DVA algorithm:

Dalsa, 605 McMurray Rd, Waterloo, Ontario, Canada N2V 2E9. (http://www.dalsa.com)

Video Cameras

Below is a selection of manufacturers of RS-170 video cameras. Many of these companies also market PAL versions of these. Whether it is RS-170 or PAL is not as important as it may seem. Almost every frame grabber card is capable of either RS-170 or PAL signal capture. PULNiX, Sony, and Watec cameras are also available through Edmund Scientific (see below).

Astrovid 2000, Adirondack Video Astronomy 26 Graves St, Glen Falls NY 12801 USA (http://www.astrovid.com)

Cohu Inc. Electronics Division, PO Box 85623, San Diego CA 92186 USA (http://www.cohu-cameras.com)

PULNiX America Inc., 1330 Orleans Dr, Sunnyvale CA 94089 USA (http://www.pulnix.com)

Sony Electronics Inc., 1 Sony Dr, Park Ridge NJ 07656 USA (http://bssc.sel.sony.com/Professional/service/index.html)

Watec America Corp., 3155 East Patrick Ln, Las Vegas, NV 89120, USA (http://www.watec.net)

Intensified Video Cameras

These are video cameras for the adventurous (as well as wealthy). Due to export restrictions, these might be difficult to obtain outside the United States.

ITT Electro-Optical Products Division, 3700 East Pontiac St, Fort Wayne, IN 46803, USA. (tel. 219-423-4341)

Night Vision Systems, 386-B Greenbrier Dr, Charlottesville, VA 22901, USA (http://users.firstva.com/nvsi/)

Frame Grabbers

The following is a list of vendors of frame grabbers known to work with Linux. There are older frame grabbers that are no longer in production (such as the DataTranslation DT2851) that could be made to work with modification of the source code and/or Linux driver. The Omnimedia Technology Inc. board is just a clone of the Matrox.,

Matrox Electronic Systems, 1055 St-Regis, Dorval (Quebec) Canada H9P 2T4 (http://www.matrox.com)

Omnimedia Technology Inc., 1800 Wyatt Dr, Suite \#12A, Santa Clara, CA 95054, USA (http://www.omt.com)

Large Scale Vendors

This is a listing of vendors that make appropriate products in more than one of the above listed categories.

Apogee, Inc., PO Box 136, Union IL 60180 USA (http://www.apogeeinc.com)

Celestron International, 2835 Columbia St, Torrance, CA 90503 USA (http://www.celestron.com)

Edmund Scientific, 101 East Gloucester Pike, Barrington, NJ 08007, USA (http://www.edmundoptics.com)

Meade Instruments Corp., 6001 Oak Canyon, Irvine, CA 92618, USA (http://www.meade.com)

Tele Vue Optics, 100 Route 59, Suffern, NY 10901, USA (http://www.televue.com)

References

1   McAlister, H.A., Hartkopf, W.I., Hutter, D.J. and Franz, O.G., 1987, *Astron. J.,* **93**, 688–723.

2   Labeyrie, A., 1970, *Astron. Astrophys.*, **6**, 85–7.
3   McAlister, H.A., 1977, *Astrophys. J.*, **215**, 159–65.
4   Coulman, C.E., 1985, *Ann. Rev. Astron. Astrophys.*, **23**, 19–57.
5   ten Brummelaar, T.A., 1992, 'Taking the Twinkle Out of the Stars', Ph.D. dissertation, University of Sydney, Sydney, Australia (see http://www.chara.gsu.edu/~theo/thesis.pdf).
6   Mason, B.D., 1994., 'Speckles and Shadow Bands', Ph.D. dissertation, Georgia State University, Atlanta.
7   Bracewell, R.N., 1999, *The Fourier Transform and its Applications,* 3rd edn, McGraw-Hill, New York.
8   Klein, M.V. and Furtak, T.E., 1986, *Optics,* 2nd edn, John Wiley, New York.
9   Gezari, D.Y., Labeyrie, A. and Stachnik, R.V., 1972, *Astrophys. J. Lett.*, **173**, L1–L5.
10 Bagnuolo, Jr., W.G., Mason, B.D., Barry, D.J., Hartkopf, W.I. and McAlister, H.A., 1992, *Astron. J.,103*, 1399–1407.
11 Johnson, H.L. and Morgan, W.W., 1953, *Astrophys. J.*, **117**, 313–52.
12 Cousins, A.W.J., 1976, *Mem. RAS,* **81**, 25–36.
13 Strömgren, B., 1956, *Vistas in Astronomy*, **2**, 1336.
14 Crawford, D.L. and Barnes J.V., 1970, *Astron. J.*, **75**, 978–98.
15 Mermilliod, J., Hauck, B. and Mermilliod, M., 1997, *Astron. Astrophys. Suppl.*, **124**, 349–52.
16 Turner, N.H., Barry, D.J. and McAlister, H.A., 1992, *Astron. Soc. Pacific Conf. Series,* **32**, 577–9 (http://pasp.phys.uvic.ca/content/contents_95_03.html#Maso).

See also: de Villiers, C. (http://www.skywatch.co.za/doublestars/speckle/index.htm).

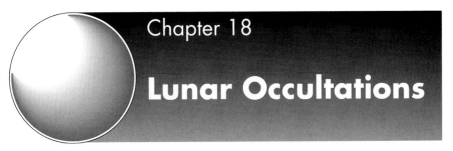

Chapter 18

Lunar Occultations

Graham Appleby

Introduction

One of the difficulties of observing and accurately measuring the relative positions and magnitudes of components of double stars is their mutual interference. Either one component is much brighter than the other, or the apparent separation between them is too small to be resolved by the optical system, particularly in the presence of distortion by the Earth's atmosphere. Ideally the components could be obscured one after the other to allow unambiguous observation of the companion as well as an estimate of the separation between them. This in essence is the principle behind application of the occultation technique to the observation of double stars.

Now, the Moon in its orbit around the Earth–Moon barycentre frequently obscures (occults) stars. As a consequence both of the inclination of the Moon's orbit with respect to the ecliptic and the precession along the ecliptic of the nodes of the Moon's orbit, all the stars in a belt of some 10° around that plane are occulted at some time during a period of about nine years. Among these are the bright stars Aldebaran, Regulus, Spica and Antares, and the star clusters Pleiades, Hyades and Praesepe. Since the Moon always moves eastward, an occulted star disappears at the Moon's eastern limb and reappears at its western limb. The phenomena can be best observed at the dark limb of the Moon, so in general disappearances are observed each month during the two weeks between New and

Full Moon, and reappearances during the following two weeks. Since the invention of the telescope, professional and amateur observers using a variety of techniques and instrumentation have recorded many thousands of timed observations of lunar occultations. Analyses of these observations have addressed such problems as improving the dynamical theory of the motion of the Moon, investigating the variable rate of rotation of the Earth, determining stellar reference frame anomalies, and measuring apparent stellar diameters and parameters in multiple star systems. It is the last two items that are of particular relevance to the subject of this chapter, but in the following sections the power of the occultation technique will be examined with reference to all of these applications.

Observation

The scientific observation of an occultation involves accurately recording the instant at which the star disappears behind or reappears from behind the lunar limb. In all but occultations of the brightest stars, telescopic or binocular aid is essential for making an accurate measurement; as the Moon approaches the star the glare from the sunlit part of the disk totally overwhelms the light from the star. By using optical aid to restrict the field of view, in most cases the star can clearly be seen at the moment of occultation.

The Moon orbits the Earth in approximately 28 days, which leads to an average easterly motion against the background of stars at a rate of 0.5" per second of time. If the instant of occultation can be estimated to a precision of 0.1 s, then the relative position of the lunar limb and the star is known at that instant to a precision of 0.05". The analysis of such observations proceeds by the computation both of the position of the centre of the Moon at that instant by interpolation in a lunar ephemeris and a precise knowledge of the position of the observer on the Earth's surface, and the position of the star taken from an appropriate star catalogue. Also, the lunar limb is not smooth; it has roughness of apparent angular extent ±2", caused by variations in the level of the lunar terrain along the line of sight from star to observer. From this information, the apparent distance of the star from the lunar limb at the instant of recorded occultation may be calculated.

Almost certainly, the computation will imply that the star should have been occulted at a slightly different time than that recorded by the observer. The reasons for the discrepancy will include errors in all the assumptions made to compute the circumstances of the occultation, such as errors in the position of the star given in the catalogue, errors in the lunar ephemeris and in the charts used to derive the level of the lunar terrain. A further correction will be attributable to the method used to make the observation. No matter how well prepared and experienced the observer, there is inevitably a time delay between the instant that the observer perceives and then records the event. If a stopwatch is used to record the event, it has been estimated[1] that this delay, or personal equation, is on average about 0.3 seconds for a disappearance and 0.5 seconds for reappearance, the larger value for the latter being due to the intrinsic "surprise" element of this type of event. Another recording technique in common use is the so-called eye-and-ear method; the observer listens to an audible one-second time signal whilst concentrating on making the observation, then mentally estimates the time of the event as a fractional part of a second. Results of analyses[1] suggest that this method is essentially free from personal equation effects, with observers achieving measurement precisions of about 0.1 seconds. A far more accurate technique is to record the occultation events electronically. A photomultiplier is used to count individual photons reaching the telescope from the star, and the counts are integrated over contiguous, short time intervals, of duration typically one millisecond. The resulting light curve can then be analysed to determine among other quantities the instant of occultation with precision close to one millisecond.

Double Stars

These then are the techniques of lunar occultation observation, where the star being occulted is a single star. If the star is in fact a double or a binary system, the intrinsic spatial resolution of the technique can be exploited to determine several useful parameters, depending upon the observing method. If the times of occultation of each of the components are measured by one of the techniques discussed above, then the

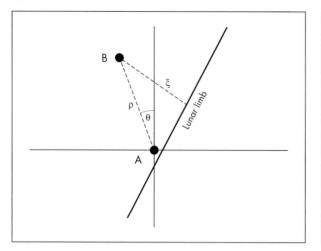

Figure 18.1.
Schematic of an occultation of a double star; the components are separated by angular distance ρ in position angle θ. The projected separation ξ may be estimated from the time difference between the two occultation events.

separation of the components, projected onto the apparent direction of motion of the Moon, can be determined simply from $\xi = \Delta t.r$, where Δt is the difference in time between the two events, and r is the rate of motion of the Moon. Now also $\xi = \rho.\cos(\theta - \varphi)$ where θ and ρ are respectively the true angular separation and position angle of the double star components, and φ is the position angle of the occultation event on the lunar limb. The situation is illustrated in Figure 18.1, where the star A is about to be occulted by the limb of the Moon. Component B of the double star is separated from component A by distance ρ in position angle θ. The distance measured by the observed time difference Δt is the projected separation ξ. Provided that the personal equation effects discussed above are the same for each of the two events, then the accuracy of determination of ξ, is limited only by the resolution of the timing technique. Of course, from observations of a single event it is not possible to determine ρ and θ. However, if a series of observations of the same double star is obtained either from different locations on the Earth or over a period of time, such that a range of values of φ is achieved, statistical methods can be used to estimate the values of ρ and θ from the deduced values of ξ. Most prediction packages that may be obtained from the references given in the Resources Section of this chapter include for each event the values of position angle θ and the rate of motion r of the lunar limb.

Visual Observations

This discussion implies that both components of the system are visually resolved during the occultation; if the components are too close together to be resolved, then the observed effect has been determined[2] to depend both on the apparent separation of the components and on their relative brightness. An analysis of a large number of occultation observations that had been made over some 35 years showed that for more than 420 of these observations the observer reported an anomalous event. The observers recorded these occultation events as not to have occurred instantaneously, to have "faded" either smoothly or in a stepwise fashion. For 160 of these events, it was found during the analysis that the 140 stars involved were in fact close doubles, many of which had been discovered by other techniques at later dates. For many of these known double and binary systems their separations and position angles were sufficiently well known to enable a calculation of the expected time intervals between the occultations of the two components, and whether the brighter or fainter component was occulted first. Intuitively it may be expected that for components of similar magnitude and for close doubles where the two occultations follow in rapid succession, the event may appear gradual, taking a slightly longer time to complete than the more normal instantaneous disappearance or reappearance. However, for wider pairs, or where the difference in magnitude of the components is large, the event might be expected to appear more dramatic, with a clear drop or step in brightness after the occultation of the first component. This expectation is borne out by the data, as shown in Figure 18.2, where for each of the observations the calculated event duration is plotted against the computed brightness-change after the occultation of the first component. The observers' comments from the original observation records have been interpreted as either "gradual" or "step" event, and these are used to code the observation symbol on the plot; circles for "gradual" and crosses for "steps". It is clearly seen that the observations are split into two classes according to whether there was a large change in brightness or long duration (step observed), or subtle change in brightness or short duration (gradual event). These results may then be used as a rough guide to interpret further visual observations of occultations, where a non-instantaneous event is observed.

The analysis discussed above concluded that a further 130 stars from Robertson's Zodiacal catalogue[3] were possibly undiscovered close doubles and would warrant closer study by say speckle interferometry, or high-speed photometric observation of future lunar occultations. At least one star, in the Praesepe cluster, from the target list given by Appleby[2] has later been confirmed as double as a result of this work.[4]

Given in Table 18.1 is a small subset of this target list showing just those stars that have been observed to fade at occultation on at least three occasions. The stars are identified by their HD, ZC and SAO numbers and visual magnitude.

Figure 18.2.
Observed event for known double stars as a function of calculated duration and brightness change; circle = "gradual", cross = "step".

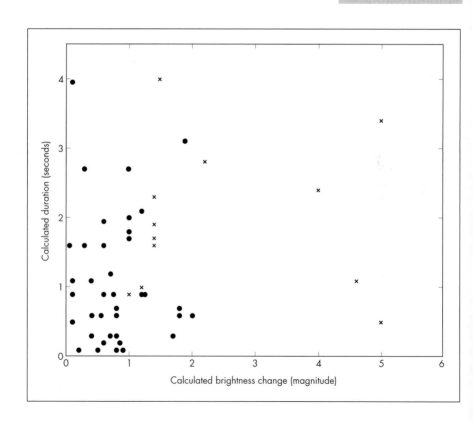

Table 18.1. List of stars that have been observed to fade on at least three occasions. For an on-line catalogue of stars which can be occulted by the Moon see the website of Paul Schlyter.[7]

HD	ZC	SAO	M_v
16302	387	75476	6.9
22017	516	93487	7.3
23288	536	76126	5.4
27934	656	76601	4.4
65736	1203	97468	7.1
88802	1500	118181	8.1
89307	1506	99049	7.1
120235	1978	139559	6.6

Photoelectric Observations

Naturally, if the photoelectric method is used to observe occultations, detection of much closer pairs should be possible. For high-speed photometry with millisecond (ms) resolution, separations of a few milliarcseconds (mas) should be detectable, far exceeding the diffraction limit of the telescope being used. However, such observations are not straightforward to analyse, since diffraction and stellar diameter effects dominate the high-resolution light curves. During an occultation event a series of alternating bright and dark fringes, the Fresnel zones, are generated and sweep across the observer during an interval of some 40 ms. The first zone, across which the intensity of the light drops smoothly to zero from a value 1.4 times its pre-occultation level, is about 13 m wide on the surface of the Earth and subtends an angle of about 8 mas at the distance of the Moon. Stars with apparent angular diameter less than about 1 mas will generate a diffraction pattern close to that expected from a point source. Those with diameters significantly greater than this will create patterns that can be considered as the sum of a series of point source diffraction patterns displaced in time relative to each other.[5] Thus for high-speed measurement of an occultation event where the diffraction pattern is sampled say at a resolution of 1ms, the characteristics of the resulting light curve will depend upon the diameter of the star. This effect is illustrated in

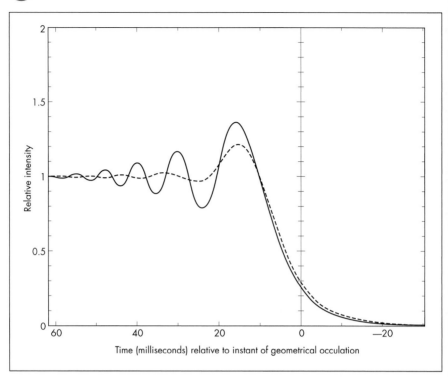

Figure 18.3 where theoretical light curves are computed for a point source and for a star of angular diameter 6 mas. The light curves illustrated here have been further modified from the purely theoretical ones to take account of a realistic bandwidth of the detector system and non-zero telescope aperture (modelled as 50 cm).

The variation apparent in Figure 18.3 of the shape of the light curve as a function of stellar angular diameter can of course be exploited in the analysis of observed light curves; both the precise time of occultation and the stellar diameter may be estimated by non-linear least squares methods. An initial estimate of the diameter is made, perhaps from previous observations or from theoretical considerations based upon the star's spectral characteristic[6] and used to compute an approximate light curve. This is then compared point-by-point with the observed light curve and the differences used to solve for corrections to the initial estimate. The process is repeated until convergence is reached and depending upon the quality and signal-to-noise ratio of the data, precisions of better than 1 mas may be achieved. In practice several other parameters are solved simultaneously with stellar diameter, such

Figure 18.3.
Theoretical light curves for occultation of a point source (dashed curve) and for a star of angular diameter 6 mas (solid curve).

as an estimate of the brightness of the star, the background noise and rate of motion of the lunar limb. A large number of stellar diameter measurements has been obtained by this method and published in the astronomical literature.

The method can readily be used for the analysis and discovery of close double stars. If evidence of duplicity is suspected in an observed light curve, the modelling process is extended in order to compute a theoretical curve by summing two such curves displaced in time and amplitude by the initial estimates of component separation and brightness and lunar limb-rate. The fitting process is identical to the single-star case, except that now two diameters may be estimated along with the parameters of the double star system. The results of such analyses are of course the same as for the visual observation method, in the sense that only the component of the double star separation in the direction of motion of the lunar limb is determined from a single observation. However, separations as small as a few mas are detectable.

Summary

The occultation technique is seen to be a valuable tool for serendipitous discovery of double stars, where visual observation can be valuable. Accurate timing of the separate events can lead to measurement of minimum separations at sub-100 mas levels of precision, as well as estimates of the relative brightness of the components. High-speed photometric observations are capable of mas-level observation.

References

1 Morrison, L.V. and Appleby, G.M., 1981, *Mon. Not. R. Astron. Soc.,* **196**, 1005.
2 Appleby, G.M., 1980, *J. Brit. Astron. Assoc,* **90**, 6.
3 Robertson, J., 1940, *Astr. Pap. Washington,* **X**, Part II.
4 McAlister, H.A., Hartkopf, W.I., Sowell, J.R., Dombrowski, E.G. and Franz, O.G., 1989, *Astron. J.,* **97**, 510.
5 Evans, D.S., 1971, *Astron. J.,* **76**, 1107.
6 Barnes, T.G. and Evans, D.S., 1976, *Mon. Not. R. Astron. Soc.,* **174**, 503.
7 Schlyter, Paul, 2002 (http://www.stjarnhimlen.se/zc/zc.html).

A great deal of information on the occultation technique may be obtained from the International Occultation Timing Association. The website can be found at http://www.lunar-occultations.com/iota/ iotandx.htm. It gives prediction information, software and links to further resources.

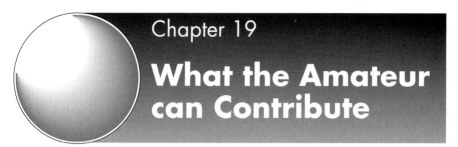

Chapter 19

What the Amateur can Contribute

Bob Argyle

Introduction

In this book we have been looking at various ways in which the relative positions and brightnesses of double stars can be measured in such a way as to contribute to the general knowledge of these objects. The main areas of opportunity can be summed up as follows.

Graticle and CCD Observations of Faint, Wide Pairs

There is no doubt that there are large numbers of relatively wide and faint systems for which little astrometry and photometry exists. The USNO have recently published on their website a list of more than 6000 pairs which have yet to be confirmed as doubles or which have been unduly neglected. These pairs are all wider than 3″ and are relatively faint but could be observed satisfactorily with a medium aperture and a graticle micrometer – see Chapter 12 and the work by Harshaw.[1] These stars would also be ideally suited to CCD astrometry even with a moderate telescope and a commercially available CCD camera. Chapter 16, written by Doug West, indicates how this can be achieved. The list of neglected pairs can also be found on the CD-ROM.

Micrometer Measures of Long Period Binaries

Many of the brightest binaries such as Castor, γ Leo and 61 Cygni have orbits graded as 4 or 5. These pairs will benefit from continual monitoring and are easy with small telescopes and micrometers. The frequency of observation should be matched to the apparent motion in the orbit, so in the case of Castor, for instance, annual means at the present time still show significant differences and should be continued for some years yet. For γ Leo, however, motion is currently very slow and means could be taken every 5 years or even 10 years without detriment. The important point is that the measures should be made since it is by no means clear that the professional community will be doing it. As techniques become more sophisticated, the close and rapid binaries are becoming the focus of attention leaving the wide visual pairs virtually unmeasured.

For pairs wider than about 10″ then most of these systems are probably optical pairs and occasional measures can serve to check on the proper motions of the component stars. It is even possible to find errors in the *Hipparcos* Catalogue, as Jean-François Courtot has done with his 21-cm reflector and filar micrometer.[2]

Relative Positions of Faint Stars from Sky Surveys

Vast amounts of untapped data on wide pairs lie in the various sky surveys taken with the world's largest Schmidt telescopes at ESO, Siding Spring and Palomar Mountain. What is more the data now encompass several wavelength bands and epochs. A determined individual, such as Domenico Gellera of Lodi, Italy, who has built and used his own measuring machine[3] can make substantial contributions because many of the pairs on these charts are not only unmeasured but uncatalogued. Sr Gellera has shown that it is possible to measure pairs as close as 5″ using a microscope fixed to a two-axis measuring machine. He has made over a thousand measures of the pairs of Pourteau and in most cases these are the first and only measures since the original catalogue was compiled from Astrographic zone plates.[4, 5] This work was done from photographic prints of the Palomar Schmidt survey and a single print typically contains hundreds of pairs. In collaboration with Willem Luyten he used his measuring machine to measure the relative positions of pairs of white dwarfs.[6]

It is not even necessary to have a measuring machine to extract data from the Sky Surveys. The USNO have created a number of large catalogues the biggest of which (the A2.0 catalogue) is the result of scanning Schmidt plates using the PMM machine at Flagstaff Station in Arizona. The result is a catalogue with 526 million stars down to magnitude 19 or so and distributed on 10 CD-ROMs. A smaller alternative is the SA2.0, with 55 million stars now only available by ftp from the USNO site.[7] UCAC1 is a more recent and more accurate catalogue based on *Tycho*-2 and USNA2.0 which contains 27 million stars between magnitudes 8 and 16 in the southern hemisphere. Pairs and multiple stars closer than 3″ are not listed. It is not quite complete covering about 80% of the southern sky. The UCAC project is continuing with the astrograph used being relocated in the northern hemisphere. The sky has now been observed as far north as +45° and the results will appear in UCAC2 in 2003 or so. UCAC1 gives positions good to 0.02″ between magnitudes 9 and 14 and 0.07″ at magnitude 16. The mean epoch is between 1998.0 and 1999.9. The data is made available in a form suitable for Unix/Linux, MAC or MS Windows.

An alternative is to use facilities such as ALADIN on the SIMBAD website. A description on how to use this facility is given by West.[8]

Visual Confirmation of Pairs in the WDS

There are several thousand pairs in the WDS which have only one observation – that of the discoverer – and the WDS project team have requested confirming observations. There is a useful opportunity here to contribute by checking these pairs and seeing, firstly if they exist, and secondly to make an estimate of the relative positions and magnitudes to see if any have moved significantly since discovery. Many of these pairs would be suitable for both visual and CCD imaging or could be located on the various sky surveys. The list can be found on the USNO website.

Photometry

Perhaps the greatest lacuna in the WDS is the lack of good photometry for many of the wider systems. With a CCD camera, it is possible to measure magnitudes for double stars in some or all of the standard wavebands such as B, V, R and I. (U can be attempted if the CCD front window is coated with a layer of ultraviolet transmitting material but this can be quite expensive.) Colours are defined such as $B-V$, $V-R$ and $R-I$ and are easily calculated from the individual magnitudes in those particular wavebands. Required filters can be made up from commercially available glass such as that made by Schott. For further information see the articles by P. Boltwood[9,10] (contact e-mail: boltwood@ fernbank.com).

Doubtless, there are many variable components yet to be discovered and in the case of double stars the great advantage is that there is a built-in comparison already available for doing differential photometry.

Lunar Occultation Observations

Graham Appleby has already described the use of lunar occultations to investigate the duplicity of previously single stars in Chapter 18. Further information on all aspects of lunar occultation work can be obtained from the International Occultation Timing Association at http://www.lunar-occultations.com/iota/iotandx.htm.

Use of Large Refractors

It is certainly true that many of the large refractors originally designed to do micrometer work on close

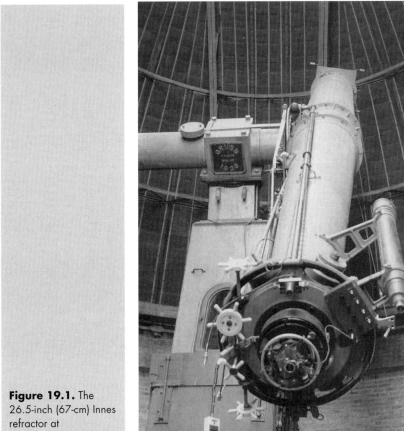

Figure 19.1. The 26.5-inch (67-cm) Innes refractor at Johannesburg, pictured in 1982 (Bob Argyle).

binaries are not currently being used for this purpose and some are almost unused, such as the great 26.5-inch refractor at Johannesburg (Figure 19.1). Some are available for research by amateur observers who have a serious programme of measurement to carry out, in particular, the 50 and 76-cm refractors at Nice, as described in Chapter 21.

Refractors of 12–15 inches in aperture, of which there are many still in working order, particularly in the USA could be employed for measuring some of the new *Hipparcos* and *Tycho* pairs. The long focal lengths of many would make them suitable for using a CCD for astrometry and photometry of faint pairs.

Calculation of Orbits

We always hope that the end product of all our hard-earned micrometer measures on a particular system will be the derivation of an orbit from the apparent ellipse and an idea of the total mass in a binary system. It may not be in our lifetime but there is a certain satisfaction from putting down a database of reliable measures that some future researcher will be able to use. Alternatively it is possible to do orbital analysis on systems which have sufficient observations to cover an arc which will allow a good estimate of the apparent ellipse to be made.

Andreas Alzner has gone into the details of orbital analysis in Chapters 7 and 8. Not only professionals, but also skilled and mathematically minded amateurs, like René Manté in France, regularly publish useful new orbital elements (cf. IAU Commission 26 Circulars). It is certainly a challenging occupation and needs a good appreciation of the problems which are posed. Now comes the awful warning. There have been some very bad orbits appearing in print. One had the companion going in the wrong direction and another used an apparent arc of 3° to calculate an orbit of several thousand years and quoted the period to one decimal place into the bargain! In an attempt to counter the proliferation of unhelpful orbits in the literature van den Bos was driven to write a paper called *Is this orbit really necessary*!

Discovery

As early as the 1840s Sir James South bemoaned the fact that F.G.W. Struve had swept the sky clear of new double stars and there was little left for him to do. Twenty or so years later when Burnham began to find many new pairs using a 6-inch telescope even T.W. Webb expressed the view that he could not hope to keep up this rate of discovery. In fact this was just the start of a golden period for visual discovery which lasted in essence until the middle of the last century. After that it is fair to say that minds were concentrated on getting more observations of the existing systems in order to accumulate stellar masses and dynamical parallaxes. Even so the work of Paul Couteau and Paul

Muller in France and Wulff Heintz in the USA indicated that there was no shortage of new pairs for those prepared to look for them with suitable apertures. The *Hipparcos* satellite which operated between 1989 and 1993 found about 15,000 new systems, some of which would have been too difficult for visual observers but some of the pairs can be resolved visually and the widest discoveries have been seen with very small telescopes. *Hipparcos*, and the associated *Tycho* mission which looked at other observations made by the satellite to a fainter magnitude but with less accuracy than the main mission, was by no means a complete survey.

In short there are still new double stars to be found either by lunar occultation or by visual examination in a concerted manner of, say, POSS films. As already mentioned, Schmidt survey films or prints can show stars down to 5″ separation. In his study of the pairs on POSS prints originally found on astrographic plates by Pourteau, Domenico Gellera noted a number of closer components in these systems. These pairs have not been confirmed so far but at typical magnitudes of 12–16 and separations of about 5″, these could be recorded with a 10-inch Schmidt–Cassegrain with a CCD camera (see Figure 16.1). The power of modern telescopes and CCD cameras is such that even pointing at a random area of sky, one is likely to record pairs which are not catalogued.

Direct visual discovery is another matter. New pairs still turn up and the French observer Jean-Claude Thorel using the 50-cm refractor at Nice has discovered four to date but these are by-products of a measurement programme rather than a deliberate attempt to survey for new discoveries. Sky conditions, particularly seeing, would need to be very good so that stars surveyed show sharp round disks and any close companion (within range of the telescope) would be relatively easily visible. It is one thing to measure a known pair whose separation is below the Rayleigh limit but it is quite another to discover one at the same distance.

References

1   Harshaw, R., 2002, *Webb Society Deep-Sky Observer*, **128**, 1–7.
2   Courtot, J.-F., 2000, *Webb Society Double Star Section Circular*, No. 9.
3   Gellera, D., 1980, *Webb Society Quarterly Journal*, **42**, 4.

4   Gellera, D., 1982, *Webb Society Double Star Section Circular*, No. 2.
5   Gellera, D., 1984, *Webb Society Double Star Section Circular*, No. 3.
6   Gellera, D., 1989, *Webb Society Double Star Section Circular*, No. 4.
7   See   http://ad.usno.navy.mil/star/star_cats_rec.html (note the underscore between star and cats and cats and rec)
8   West, D., 2001, *Webb Society Double Star Section Circular*, No.10.
9   Boltwood,   P.,   Dec.   1991–Feb   1992,   *IAPPP Communications*, No. 46, 48–56.
10   Boltwood, P., 1996, *IAPPP Communications*, No. 62, 45–6.

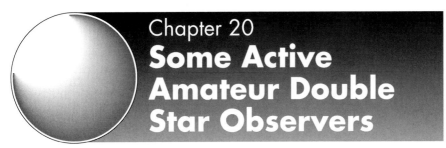

Chapter 20
Some Active Amateur Double Star Observers

Bob Argyle

Introduction

Although some effort has been expended to try and collect as much information as possible about the current activities of individuals and groups involved in double star observing the following notes should be taken as a guide only. In each case the contact details are given in the Appendix.

USA

The Celestron Micro Guide project is an international collaboration of eight observers (three from the USA, two from the UK, three from Spain), with the goal of determining whether or not the MicroGuide eyepiece (see Chapter 12) has enough sophistication, in the hands of an experienced observer, to yield useful scientific measures of separation and position angle. The members agreed to work under a protocol that calls for each to measure a few "fixed" pairs from a list of about 125 culled from the *Washington Double Star Catalogue*. The purpose of this preliminary set of measures was to determine (a) whether or not the observer has sufficient skill, (b) whether or not the MicroGuide can produce reliable results when compared to known pairs, and (c) establish the "scale constant" of the linear scale for each participant, as this value will vary from telescope to telescope and whether or not a Barlow lens is used.

The results of this work were published in *The Deep-Sky Observer* of the Webb Society. Overall, the results from eight observers showed that it was possible to achieve ± 1° in PA and ±1″ in separation. The average value for the angular size of a single division was 14″.

Also in the USA is the *Double Star Observer*, an international journal that is devoted exclusively to visual double star astronomy, edited and produced by Ronald Tanguay. The aim of the magazine is to encourage amateur/professional cooperation in the field of double star astronomy, and to provide amateur visual double star observers with easy access to a journal where they may publish the results of their research and observations.

Argentina

Mauro Gallo, from Buenos Aires, reports that members of the Amigos de la Astronomia group have begun observing and measuring binary stars using a 10-inch Meade LX-200 telescope with an ST-7E CCD camera for the imaging and using the Astrometrica program with

Figure 20.1. An image of the pair WDS18510–1747 =HJ 2832 (θ = 338°, ρ = 28″.7, magnitude A = 8.7, magnitude B = 10.6). This image was taken on 2001, Aug 28 at 23:56:28 UT. Exposure time: 10 seconds. Camera cooled to –20 °C. The A star of this pair is the brightest star in the centre of the frame image whilst star B is up and to the right to the A star. There are 228 reference stars in this unusually rich field. Note the epistence of other pairs on this image, none of which has been catalogued as yet (M. Gallo).

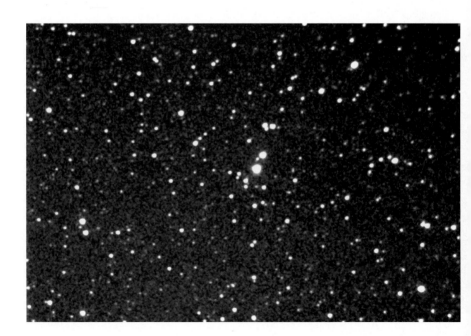

the USNO A2.0 catalogue (reference stars) to carry out the reductions. With this equipment they are able to measure binaries with magnitudes down to $V = 17$ and in the zone +30 to –90° declination, with separations from 5″ upwards with high precision. Fig. 20.1 shows an example of their work.

France

France has always been a centre of excellence for double star studies. In the last century observers such as Robert Jonckheere and Paul Muller were very active observers and discoverers. The latter also developed the double-image micrometer. The leading amateur was Paul Baize who was not only a prodigious observer but also computed orbits, many of which remain in the catalogue today. Antoine Labeyrie developed speckle interferometry which has had a profound effect on the observation of very close visual binaries and which has allowed large telescopes to be used to their full resolution capability.

For the present generation, the leading professional figure is undoubtedly Paul Couteau (Figure 20.2) with more than 2700 discoveries to his credit and 25,500 measures. Dr Couteau has spent a great deal of his career at the Observatory of Nice where today double star research still continues.

Figure 20.2. Dr Paul Couteau (right) with Bob Argyle at Santiago de Compostela in August 1996 (Angela Argyle).

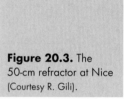

Figure 20.3. The 50-cm refractor at Nice (Courtesy R. Gili).

Under the auspices of the Commission des Etoiles Doubles of the Société Astronomique de France, a team composed of Guy Morlet, Maurice Salaman and René Gili has for some years now been taking advantage of the capabilities of the CCD imaging technique using the 50- and 76-cm refractors at Nice Observatory (Figures 20.3 and 20.4).

Whilst the 17.89 m focal length of the 76-cm refractor did not require any change, the 7.50 m focal length of the 50-cm refractor has been brought to 15.50 m using a 2× Barlow lens (Clavé). The CCD camera presently in use is a French LE2IM, a Hi-SIS 23 with a Kodak matrix KAF 401E (758 × 512 square pixels of 9 μm).

The imaging software is either QMIPS 32 or QMIPS. Short exposures of 1 s down to 0.02 s are taken. For

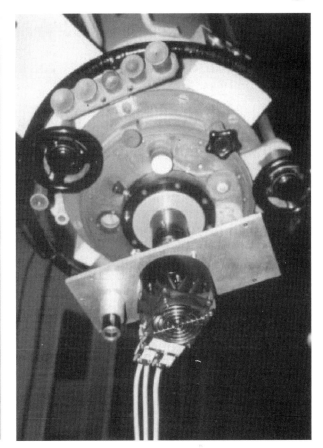

Figure 20.4. The plate at the back of the refractor can be shifted in the focal plane. It supports both the CCD camera and the eyepiece used for visual control of the field (Courtesy R. Gili).

every pair, 200 images or so are currently saved on the hard disk of a portable computer.

Observations are later reduced after the 10 or 15 best images have been selected and composited (i.e. shifted and added) using MIPS. The measurement of composite images is achieved using specific software for determining the position angle, angular separation and magnitude differences.

From 1997 to 2000, seven observing sessions have been conducted at Nice Observatory and the team measured some 300 different pairs down to 0''.4 with the 50-cm refractor and to 0''.3 with the 76-cm refractor, demonstrating that the CCD imaging technique fits the needs of double star measurement well, giving very reliable results and allowing the best use of observing time.

Jean-Claude Thorel (Figure 20.5) is one of the leading visual observers in France today. His interest in

Figure 20.5. Jean-Claude Thorel in his office at Nice. (Courtesy J.-C. Thorel)

astronomy started during a childhood illness when he was kept in isolation and his father brought him a book on astronomy to pass the time. It was some 15 years later that the interest in astronomy returned and he bought a 60-mm refractor to use at his home in Villepreux, close to Versailles. This was followed by a 20-cm Schmidt–Cassegrain and his early interests included lunar and planetary drawing and deep-sky observation. His first serious work was comet observation, resulting in a published guide on to how to observe and draw them.

He then became involved in work to resolve some inconsistencies in double star catalogues during the construction of the *Hipparcos Input Catalogue*. This involved two trips to use the 1-metre telescope at Pic du Midi in 1986 and 1987. This expanded into a general programme to measure neglected and problem pairs in the double star catalogues using the 50-cm and 76-cm refractors at Nice. He has recently been working on a programme of checking the double stars discovered by the *Tycho* mission on the *Hipparcos* satellite, some 4800 of which are visible from Nice. This had meant travelling from Villepreux to Nice three or four times a year, a return trip of 2,000 km but his job now means that he is able to live in Nice and take advantage of the proximity of the telescopes there.

He has made 6000 micrometric mean measures with the refractors at Nice, including four new pairs (JCT1-4) and has also published a biography of Robert Jonckheere amongst other works.

Figure 20.6. The 205-mm reflector used by Jean-François Courtot in Chaumont, north-east France (Courtesy: J.-F. Courtot).

Meanwhile in North-East France, Jean-François Courtot has been engaged in double star research since 1993 but he has been interested in astronomy from youth. He uses a homemade 205-mm Newtonian from Chaumont (Figures 20.6 and 20.7).

For wide pairs, a chronometric method, the transit method, is often used. The angular separation is derived from the time needed by components to successively cross the same thread because of diurnal motion. Each measurement consists of six alternate readings (±180°) of the position angle and 20 determinations of the transit time. The mean internal error for the position angle is usually ± 0°.2 and ± 0″.3 for the angular separation.

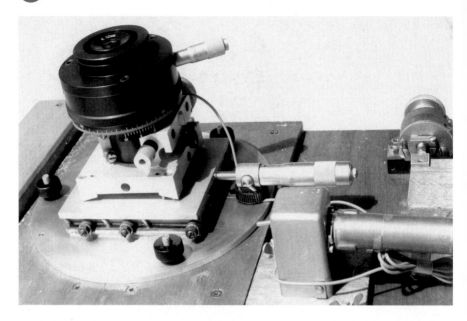

For closer pairs, a filar micrometer has been installed to measure separations occasionally down to 0″.66, the practical diffraction limit under good seeing with the 205-mm telescope. Each measurement consists also of six alternate readings of the position angle while three double-distance measures of separation are taken. For pairs close enough to be observed at the same glance under magnification ×500 without darting rapidly from one star to the other, the filar micrometer allows the mean internal error to be kept typically within ±0°.1 and ±0″.03. This latter limit is the equivalent reading accuracy allowed by the screw constant and the overall focal length.

Figure 20.7. The RETEL micrometer attached to the 205-mm reflector of Jean-François Courtot (J.-F. Courtot).

To compensate for various seeing conditions and more or less controllable errors, the measurement of a given double is usually repeated on 3 or 4 different evenings. For the closest pairs, bright components ($V \leq 7.5$) and stable seeing are needed. Wide pairs accommodate to fair conditions and can sometimes be measured down $V = 10$.

So far, some 3000 measurements of 800 different doubles have been completed, published and included in the WDS database, and a few of these pairs having never been observed before. Aside from observations of orbital and neglected systems, proper motions of optical pairs are checked using historic double star measurements as a start point and new determinations are proposed at times.

Germany

The leading observer is Andreas Alzner who operates a 32.5-cm Cassegrain (Figure 20.8) and 35-cm Newtonian in an observatory at Hemhofen, just outside Erlangen. The telescopes are equipped with both RETEL and van Slyke filar micrometers and a Méca-Précis double-image micrometer.

Dr. Alzner has also published a number of orbits in *Astronomy and Astrophysics* and concentrates on close

Figure 20.8. The 32.5-cm Cassegrain of Andreas Alzner fitted with a double image micrometer.

pairs, down to 0.25″, including also some of the first measures of *Hipparcos* discoveries from the ground.

Hungary

The Hungarian Double Star Section, established in 1992, publishes a column in the monthly journal of the Hungarian Astronomical Association, *Meteor* and is led by Tamás Ladányi.

There is also a double star circular, *Binary*, which is published once a year. It includes articles, translations and maps about double stars and is edited by Tamás Ladányi.

Between 1991 and 2000 about 7000 mainly visual observations were made but several observers are now making measures, in particular Ernö Berkó who has made more than 200 measures with his CCD camera and 35.5-cm reflector, Andras Dan using a micrometer and a 20-cm Maksutov–Cassegrain reflector and Tamás Ladányi who measured doubles in 1999 with the 50-cm refractor in Nice, and has also made measures at the University of Szeged using a Celestron-11 and CCD camera.

New Zealand

In the southern hemisphere, very few observers are active, even though there are many underobserved systems. One exception is Ormond Warren whose interest began in primary school when the headmaster allowed him to use his 3-inch refractor.

In 1987 he became interested in double star astronomy after moving to Wanganui, where he found the city observatory had a fine history in this field. His first study was the set of pairs and triples originally discovered at this site, early last century, and given the discovery designation NZO. Fortunately he was able to use the famous 23-cm f/15 Fletcher equatorial (by Cooke the Elder, 1860) (Figure 20.9) and an antique Cooke & Sons Type-A (English) bifilar micrometer, the same instruments used by the discoverers. By 1990 this project expanded to a general survey of southern hemisphere pairs and multiples.

Figure 20.9. The Fletcher equatorial at the Ward Observatory, Wanganui. Reproduced by kind permission of Wanganui Astronomical Society.

Currently he is a planetarium presenter at the Stardome Observatory and Planetarium in Auckland city, where he also undertakes a measuring programme. His most recent papers were published in the Royal Astronomical Society of New Zealand (RASNZ) journal *Southern Stars* in 2000 (two) and 2001 (one). Ormond is acting as Observation Coordinator and has produced a number of guides to various aspects of observing. In addition he has produced a number of lists of pairs which need to be observed for signs of change in relative position since last measured. Some effort has been made to scrutinise and make approximate measures of these pairs. Many of them have been measured only once (at discovery), and hence confirming observations are necessary. About 30% of those checked so far have been found to be discordant with the data in the *Washington Catalogue of Visual Double Stars* (WDS). These "anomalies" are being compiled into a list for further study: they involve pairs with erroneous positions, erroneous magnitudes perhaps due to variability, and some which at present cannot yet be identified in a field with other nearby pairs. Some of the pairs have changed to such a degree that mere visual scrutiny would alert the observer to the fact that motion had taken place. In most of these cases one of the stars involved has a relatively high proper motion rather than the pair being a true binary.

The Double Star Section of the RASNZ was formed in April 2000 and is currently headed by the Director, Warren Kissling. Its purpose is to observe and measure doubles in the southern hemisphere, many of which suffer from a lack of regular observation. Two of the members of the section have RETEL micrometers and are learning how to make useful measures with them.

Of particular interest are the NZO pairs, already mentioned. Another member, comet discoverer Rod Austin, has been working to produce a definitive set of these pairs that Ormond Warren and he have worked on over the past 10 years or so.

One of the main problems for the future may well be to encourage more observers to take on a serious programme of measurement, but one advantage of being in New Zealand is access to relatively clear, dark and (in terms of double-star work) neglected skies. There is, however, a dearth of measuring equipment which, partly due to the low exchange rate for NZ dollars, tends to be very expensive.

Spain

Over the last 40 years, a small group of Spanish amateurs has been systematically measuring visual double stars. They are currently preparing to publish all the measurements made between 1970 and 2001 – some 10,000 in all. This massive work was presented in October, 2000 at the annual meeting of the Astronomical Society of France's Double Star Commission, held in Castelldefels, near Barcelona, Spain. An Internet version will soon be available.

The first measurement catalogue entirely produced in Spain by an amateur was that by José-Luis Comellas (a lecturer in Modern History at the University of Sevilla – see Figure 20.10). The first, published in 1973 (*Catálogo de Estrellas Dobles Visuales 1973.0*), contained measurements of 1200 double stars, using a simple micrometer and a 75-mm aperture Polarex-Unitron refractor. Twelve years later, the same author

Figure 20.10. The Spanish double star observer, José-Luis Comellas (T. Tobal).

published a second catalogue (*Catálogo de Estrellas Dobles Visuales 1980.0*) that included 5104 doubles within reach of his new 102 mm aperture Polarex-Unitron refractor, of which he personally measured over 3500.

Since 1985 other observers have maintained the continuity of Comellas' work. From 1976, Tófol Tobal regularly collaborated with him, and in the mid-1980s he built a small observatory equipped with a 102-mm Polarex Unitron refractor and a filar micrometer, allowing him to start a systematic revision and update of the 1980.0 catalogue. In 1991, in conjunction with other colleagues, Mr Tobal coordinated the measurements sent by individual observers and began to publish a circular (*RHO: Circular de Estrellas Dobles Visuales*) for internal use, in order to coordinate the work and to publicise the results. Recent acquisition of new precision micrometers, double image Lyot-Camichel-like, and CCD devices have been made, and between 1992 and 2000 more than 5000 new observations and measurements has been collected, provided by amateurs throughout Spain.

In 1991 the Garraf Astronomical Observatory (OAG) was founded and on the original site (1992–1998) it had a

3.5-metre diameter dome with a 260-mm aperture Newtonian. A new observatory has been constructed (Figure 20.11); using public and private investment and is located 30 km south of Barcelona, inside the Garraf Natural Park and was opened in November 2001. It has a new 3.5-metre dome and a 30-cm Newtonian–Cassegrain f/3.5 and f/12 telescope fitted with a CCD camera and a Lyot double-image micrometer.

Figure 20.11. The observatory in the Garraf National Park, near Barcelona. (T. Tobal)

South Africa

The Double Star Section of the Astronomical Society of Southern Africa is led by Chris de Villiers. He has recently successfully experimented with speckle imaging using the 18-inch refractor at the South African Astronomical Observatory in Cape Town. More details can be obtained from his website which is given in the appendix.

United Kingdom

The Webb Society Double Star Section started in 1968 and Bob Argyle became Director in 1970. It was not

Figure 20.12. The joint meeting between Spanish and French double star observers held in October 2000. (R. Casas).

until the end of the decade that some preliminary attempts to measure double stars using grating micrometers and home-made filar micrometers was made. By the end of the 1980s the availability of commercially made filar micrometers allowed members to make micrometric measures. At the time of writing the results have been published in eleven *Double Star Section Circulars* most of which have now been incorporated in the Observations Catalogue of the United States Naval Observatory. Using the 8-inch refractor at the Cambridge Observatories, Bob Argyle is carrying out a programme of visual measurement (see Chapter 21). The programme consists of a number of long-period binaries plus observations of some wider, fainter pairs which have not been observed for some time. Some 4300 measures have been made since 1990.

Tom Teague, using an 8.5-inch reflector near Chester, has developed a new and more efficient way of using a Celestron Micro Guide eyepiece and he is currently using it as part of the assessment programme which is described by Richard Harshaw (USA) earlier in this chapter.

Martin Nicholson operates a 12-inch Meade LX-200 telescope and a SBIG ST-7E CCD camera at Daventry in Northamptonshire. He has made numerous measures of neglected double stars from the USNO lists and has observed and measured a number of previously uncatalogued pairs using the Space Telescope Science Institute on-line Schmidt catalogue. He can measure as many as 20 pairs per hour down to magnitude 16 and with separations down to 5″. His results appear in *the Webb Society Double Star Section Circulars* and on his website.

Further Reading

Argyle, R.W., Courtot, J.-F. et al., 1983–2003, *Webb Society Double Star Section Circulars*, Nos. 2– 11.

Courtot, J.-F., 1995, *Observations et Travaux, S.A.F.,* **44,** 22–41.

Courtot, J.-F., 1996, *Observations et Travaux, S.A.F.,* **47,** 47–74.

Courtot, J.-F., 1999, *Webb Society Deep-Sky Observer,* **119,** 4.

Harshaw, R., 2002, *Webb Society Deep-Sky Observer,* **128,** 1.

Morlet, G., Salaman, M. and Gili, R., 2000, *Astron. Astrophys. Suppl.,* **145,** 67.

Nicholson, M., 2003, *Webb Society Double Star Section Circular No. 11.*

Nicholson, M. (http://www.double-star.org)

Salaman, M., Morlet, G. and Gili, R., 1999, *A&AS,* **135,** 499.

Warren, O., 2000, *Southern Stars,* **38,** 217; **39,** 9; **40,** 19

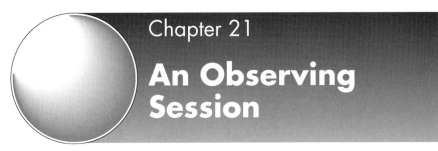

Chapter 21

An Observing Session

Bob Argyle

The Telescope

In this chapter I describe a typical observing session with the 8-inch (20-cm) Thorrowgood refractor at the Institute of Astronomy in Cambridge. The telescope belongs to the Royal Astronomical Society but is on permanent loan to the Cambridge University Astronomical Society and has been on its present site since 1930 (Figure 21.1).

It was built by Cooke in 1864 for the Reverend W.R. Dawes who did not have much opportunity to use it. It passed through the hands of W.H. Maw, a founder member of the British Astronomical Association and an active double star observer, before ending up in the possession of W.J. Thorrowgood, who, in turn, bequeathed it to the RAS.

The telescope is on a German mount and is driven in RA by a small synchronous electric motor. The focal length of the object glass is 114 inches giving the telescope a focal ratio of just over f/14 and a scale at prime focus of 71.2″ per mm. There are slow motion controls in both RA and Dec each of which run on tangent arms and consequently have to be reset every night or two. The telescope can be used either side of the pier but my own practice is to work on the east side of the pier since clamping the telescope this side is much easier and speeds up observing. In addition, the slow motion controls are to the right of the eyepiece and are more comfortable to work with.

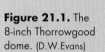

Figure 21.1. The 8-inch Thorrowgood dome. (D.W.Evans)

The Micrometer

I use a RETEL micrometer to make the measures (Figures 21.2 and 21.3). There are three wires in the field of view of the eyepiece. Two are fixed and perpendicular to each other; the third moves in two directions and, used in conjunction with the fixed wire parallel to it, measures the separations whilst the other wire is used for position angles. The movable wire is controlled by an engineering micrometer screw which has a range of about 11.5 mm and which can be read to 1 micron using the fitted vernier. The wires have a diameter of 12 microns, which translates to 0″.85 in the focal plane. As the telescope will resolve pairs about 0″.55 apart this is plainly unsatisfactory. This can be easily overcome by means of a Barlow lens. In this case I employ a ×3 Barlow which triples the effective focal length and reduces the apparent size of the wires in the eyepiece to about 0.3″. In conjunction with the 18 mm Kellner eyepiece supplied with the micrometer this gives a magnification of about ×450 and this is used for all measures.

Figure 21.2. The RETEL micrometer and its illumination power supply. (R. Sword (IOA)).

The field is illuminated by a single red LED which can lead to parallax problems if the illumination is set too high. A way out of this is to locate an LED in the telescope dewcap thus illuminating the field more evenly. On bright stars it is best to turn the illumination down or even off to set the wires since they can be seen in shadow against the star disks. Using the manufacturer's illumination I can measure wider pairs down to about V=10 and for faint, close pairs then STF 1280 (magnitudes 8.9 and 9.1 at 1.2″) represents the limit for the 8-inch refractor.

Although it is clearly better to have a micrometer residing permanently on the telescope, in my own case this is not possible since the telescope is often used for other observations including solar projection. Hence it must be fitted and removed for each observing session. I therefore have to check the instrumental position

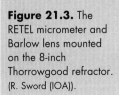

Figure 21.3. The RETEL micrometer and Barlow lens mounted on the 8-inch Thorrowgood refractor. (R. Sword (IOA)).

angle of standard pairs at the beginning and end of the night. I also measure the separations of the same pairs to give a determination of the scale of the micrometer by taking a mean of the two determinations which usually agree to within 1%.

Whilst the micrometer is being fitted to the telescope, the dome is opened to allow the inside air to come to the same temperature as the air outside. As the dome is fairly small this does not take very long. A note is made of the dome temperature at the beginning and end in case refraction corrections need to be made and to check whether any scale variation in the micrometer with temperature is discernable. In practice I don't do this. For pairs < 30″ in separation the correction is very small.

Other Accessories

For observing I take the following items:

- A notebook in which the raw micrometer readings are written. These are transcribed to another volume later and the reductions made at home.
- A star atlas with the target stars marked. I have found that using *Norton's Star Atlas* and simple star-hopping is adequate in the vast majority of cases. The telescope is fitted with setting circles but there is no sidereal clock in the dome and the circles are not that easy to read in subdued light. When the pair is in a rich starfield it is occasionally necessary to take a more detailed map of the stars nearby in order to locate the pair in question.
- A torch with a piece of red plastic over the window allows the micrometer readings to be clearly seen as well as affording enough light to write down settings.
- Finally a list contains the stars to be measured along with the number of nights which each one requires and the number left to do.

Measurement Plan

My policy is to measure the most interesting binaries on at least five nights each year. Some such as Castor (alpha Gem) tend to get more than this because the star is so bright it can be seen in twilight and observing can start earlier when the seeing can be rather good. The standard pairs used also tend to be rather bright for the same reason. Relatively close pairs (at around 1″) which are measured occasionally because they are slow-moving get four nights and any other pairs (usually wide) get three nights. As for the number of settings made on each individual star this tends to depend on the difficulty of the pair. In the summer of 1999, for instance, the fine binary zeta Her which consists of stars of magnitude 2.9 and 5.8 was separated by just under an arcsecond. This meant that measuring the companion depended very much on sufficiently good seeing but, even so, setting the position angle wire resulted in values which scattered by as much as 15 or 20°. In this case, I make up to eight settings in position angle. For wider pairs, where the separation is perhaps 20 or 30″, the agreement between individual angle settings is usually better than one degree and four measures are deemed sufficient.

It is very useful to mark up the target stars on the star atlas because another time-consuming activity is moving the dome by hand. By concentrating on a number of pairs in the same region of sky not only can these be observed more quickly but a comfortable observing position need not be disturbed too often. Having said that, trying to see stars near the zenith with a long-focus refractor requires the ability of a contortionist and I tend to avoid stars which are too high in the sky. There is no doubt that comfort is a significant advantage in securing better measures.

The pairs to be measured will depend on several factors, the prime one being the seeing. If the seeing turns out to be particularly good then I tend to concentrate on the closest pairs. If seeing is poor then wider pairs can be tried. It is very rare in Cambridge that stars of 1″ separation cannot be measured so it is clear that the city environment is not necessarily a bad one even though the sky is usually rather bright. Another factor may be the number of observations left for a particular pair. It is better although not necessary to try and get sufficient measures for a mean during the same season. For wider pairs which are slow moving it may be three or four years before I get sufficient measures for a mean.

A red torch is used throughout: for examining the star atlas for the location of the next pair, looking at the verniers on the micrometer and writing down the settings in the observing sheet. A simple hand-held torch with a button to allow the light to be flashed on and off is most efficient. Rechargeable batteries soon recoup the initial outlay.

Measurement

In measuring each pair the position angle is always done first and although formally the wire should be reset at the end of this procedure to the mean value in practice this is not done since the individual values tend to agree closely enough for this purpose. Two to four settings are made and the individual angles remembered before writing them down. For wide pairs these will usually agree to within one degree and it is then only necessary to remember the decimal part. It is recommended that the quadrant in which the fainter star lies is noted. With equatorial telescopes the approximate directions of the cardinal points are usually fairly obvious so it is a simple matter to record whether the companion star is in the first quadrant (i.e. with a PA between 0 and 90°) or another quadrant. This is because the recorded PA from the micrometer is ambiguous by 180° depending on where the microm-eter barrel is pointing. I happen to be right-handed so the micrometer barrel is usually in the first or second quadrant.

For separation, the technique used depends on the distance between the stars. For close pairs (< 15″) the double distance method is used and the two values of the screw are written down at the end of the proce-dure. For wider pairs it is too time-consuming to do this so four settings are made with the movable wire on one side of the fixed wire then another four set-tings made with the movable wire on the opposite of the fixed wire. This requires the use of the telescope slow motions and this is where a box screw would be useful. On the older brass micrometers this was an arrangement which allowed all the wires to be moved across the field of view whilst retaining their absolute position with respect to one another. With the RETEL micrometer the separation readings are in mm on the micrometer screw but each revolution of the screw is graduated in 50 divisions so care must be taken to note whether the reading is between x and $x + 0.5$ mm

Figure 21.4. An extract from the author's observing book. The two central columns record the settings of the movable wire in millimetres corresponding to the double distance method. The right hand column gives the observed position angle on the micrometer barrel. This is converted to the true PA and separation by using the reference pair δ Boo. The final observed PA and separation are given along with the epoch of observation in decimals of a year. Note the correction to the mean PA of STF 1932. It is easy to misread the micrometer dials in the dome!

or $x + 0.5$ and $x + 1.0$ mm where x is the reading in whole millimetres on the barrel. In most cases, however, the error will stand out easily and be corrected when reducing the data.

As mentioned above for the Thorrowgood it is necessary to remove the micrometer and Barlow assembly at the end of each session and so one of the first pairs to be measured is a calibration pair. A list of bright pairs with separations from 14 to 100″ around the sky is used (and is given in Chapter 15). The relative position angles and separations are known to about $0°.1$ and about 0.05″ – sufficiently small to be negligible compared to measurement or personal errors. The same pair, if possible is also measured at the end of the night. If it is possible to leave a micrometer in place on the telescope then this is the best option – even so, the zero of position angle should be checked at least once per night.

Reducing Observations

The observed micrometer settings are taken home where they are copied with a little more neatness into an observing book (Figure 21.4). The original

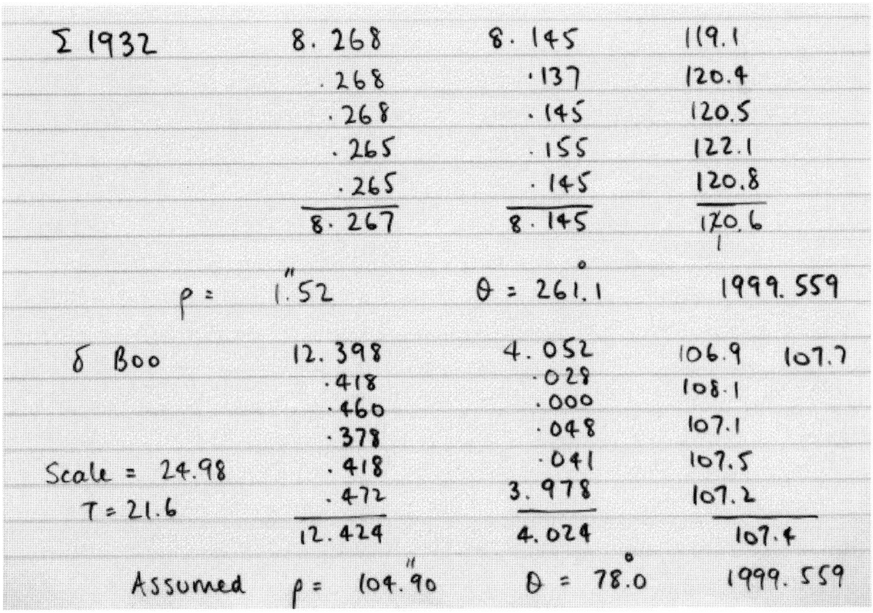

Σ 1932	8.268	8.145	119.1
	.268	.137	120.4
	.268	.145	120.5
	.265	.155	122.1
	.265	.145	120.8
	8.267	8.145	120.6
$\rho =$ 1″.52		$\theta = 261.1°$	1999.559
δ Boo	12.398	4.052	106.9 107.7
	.418	.028	108.1
	.460	.000	
	.378	.048	107.1
Scale = 24.98	.418	.041	107.5
T = 21.6	.472	3.978	107.2
	12.424	4.024	107.4
Assumed $\rho =$ 104″.90		$\theta = 78.0°$	1999.559

recordings are kept in case of a query or transcription error. It is at this point that the mean settings are calculated and the position angles and separations worked out.

The two observations of the calibrations are done first. This gives a mean value for the observed position angle at the beginning and end of the session. This usually agrees to better than 1 degree. The difference between the instrumental value and the value from the calibration list is the correction to be applied to all the other mean position angles. Similarly a mean screw value is obtained from the calibrations and applied to the remaining observations. The final touch is to convert the calendar date to a decimal of a year. This can be done via a lookup table which can be found in the Explanatory Supplement to the Astronomical Ephemeris or the program JD&Epoch in the "soft" folder on the accompanying CD-ROM can be used. High-resolution work such as speckle interferometry on rapid visual binaries demands using the date to four decimal places but for visual work with small telescopes, three places of decimals is more than adequate.

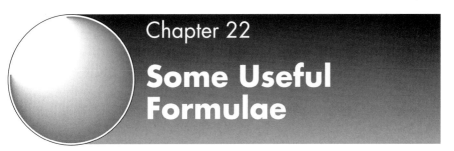

Chapter 22

Some Useful Formulae

Michael Greaney

Introduction

The observations brought inside after a night at the telescope represent just raw data. A number of steps must be taken to reduce these data to meaningful observations. These steps will include expressing the time of each observation as a standard epoch and reducing the observed magnitudes of the individual components. Consideration will also have to be given to any effects that atmospheric refraction might have on the relative positions of the components. If the observations are to be reduced to some standard epoch then corrections must be made for the effects of precession and proper motion on the position angle.

Dating Observations

The date of a double star observation should be expressed as the year in fractional form, usually to three or four decimal places. This is known as the epoch of the observation. There are two forms of epoch: the Besselian epoch and the Julian epoch.

The Besselian epoch is based on the length of the Besselian year of approximately 365.2422 days and is given by

$$\text{Besselian epoch} = \text{B}1900 + (\text{JD} - 2415020.31352) / 365.242198781$$

where the prefix B indicates that it is a Besselian epoch, JD is the Julian date and the constant 2415020.31352 is the Julian date of the standard epoch B1900, i.e. 1900 January 0 (= 1899 December 31).

The Julian epoch was introduced with the new astronomical constants in 1984. It is based on the length of the Julian year of exactly 365.25 days and is given by

$$\text{Julian epoch} = \text{J2000} + (\text{JD} - 2451545) / 365.25$$

where the prefix J indicates that the epoch is a Julian epoch and the constant 2451,545 is the Julian date of the standard epoch J2000, i.e. 2000 January 1 at 12 hours Universal Time (UT).

The prefixes B and J are used only where context or accuracy make them necessary.

The Besselian epoch is the one normally used for dating double stars observations and is quoted to three decimal places for visual observations. This means effectively dating each observation to an accuracy of nearly nine hours, so a single epoch value could serve for a whole observing session.

The Julian date (JD) can be found from the following algorithm, where year, month, day and hour refer of course to the time the observation was made.

IF month > 2 THEN ym = year + 1 ELSE ym = year
L = INT(7 * ym / 4) + INT(3 * INT((ym + 99) / 100) / 4)
J0 = day + INT(275 * month / 9) + 367 * year –
    L + 1721028.5
JD = J0 + hour / 24

INT means the "integer part of ", or more precisely, the highest integer less than the number, e.g. INT(5.6) = 5, but INT(–5.6) = 6.

The hour is expressed in decimal form, i.e.

$$\text{hour} = \text{hours} + \text{minutes} / 60 + \text{seconds} / 3600$$

so if the time were 21:22:30 then the hour would be 21.375.

J0 is the Julian date at 0 hours UT. Dividing the hour by 24 and adding it to J0 gives JD.

As an example the Julian date and the two epochs for 9 pm UT on Christmas Day 2003 would be:

Julian date = 2452999.375
Besselian epoch = 2003.9832
Julian epoch = 2003.9819.

The Julian date for which both the Besselian and Julian epochs have the same value is 2429698.882870183, i.e. 1940 March 10 at 9:11:20 UT.

Precessing the Position Angle

When the position angle measurement is made it is made with respect to the north celestial pole at the time of the observation. The pole, however, is not a fixed point in the sky by varies due to the effects of precession. This means that the position angle can change over time without any orbital motion of the companion star. Furthermore, the proper motion of the star causes the star to change its position with respect to the pole and consequently induces changes in the position angle.

It is important, then, to reduce the observed position angle to some standard epoch so that any changes observed in the position angle will be due to orbital motion and not to the variability of the reference frame.

Likewise, when calculating the position angle and separation from the orbital elements (see Chapter 7) it is important to allow for these effects. The table of orbital elements should give the epoch of the position angle of the ascending node (Ω), but unfortunately not all tables of orbital elements give it. When it is given there are two ways to carry out the calculation. First calculate the position angle and separation and then reduce the position angle to the date of observation, or first reduce the position angle of the ascending node to the date of observation and then calculate the position angle and separation. The result is the same either way.

Proper Motion

The first step in reducing the position angle – either the observed or calculated position angle – is to apply the correction for proper motion. The change in position angle due to proper motion is given by

$$\Delta\theta_\mu = -0\overset{s}{.}00417\mu_\alpha \, \sin d \, (t - t_0)$$

where $\Delta\theta_\mu$ is the change in position angle due to proper motion,

μ_α is the proper motion in right ascension in seconds of time per year,

δ is the declination of the star,

t is the epoch of observation, and

t_0 is the epoch to which the position angle is being referred.

The constant converts seconds of time (of the proper motion in right ascension) to degrees.

Precession

The changes due to precession are then applied. These can be found from the formula

$$\Delta\theta p = -0°.0056 \sin \alpha \sec \delta (t - t_0)$$

where $\Delta\theta p$ is the change in position angle due to precession α is the right ascension of the star.

This is an approximate formula, however, and should not be used for stars that are close to the pole. A rigorous formula, according to Green[1], is given by

$$\tan(\phi - \phi_0) = \frac{\sin(\alpha - z_A)\sin \theta_A}{\cos \delta \cos \theta_A + \sin \delta \sin \theta_A \cos(\alpha - z_A)}$$

where ϕ is the position angle referred to the equator and equinox of date and ϕ_0 is the position angle referred to the standard equator and equinox (i.e. $\Delta\theta_p = \phi - \phi_0$). θ_A and z_A are precessional angles which, for the standard epoch of J2000, are given by

$$z_A = 0°.640616T + 0°.0003041T^2 + 0°.0000051T^3$$
$$\theta_A = 0°.5567530T + 0°.0001185T^2 + 0°.0000116T^3$$

where T is the interval $(t - t_0)$ expressed in Julian centuries of 365,25 days.

In terms of the precessional angles, the approximate formula could be written as

$$\Delta\theta_p = -\theta_A \sin \alpha \sec \delta.$$

Reducing the Position Angle

The total change in the position angle due to the effects of proper motion and precession is simply

$$\Delta\theta = \Delta\theta_\mu + \Delta\theta_p.$$

The effects are most pronounced for stars of high declinations. It is possible for these effects to cancel out by being of equal value but of opposite signs.

The reduced position angle, θ_0 is just

$$\theta_0 = \theta + \Delta\theta$$

Example

The position angle for alpha Centauri for the year 2000 is 222°3. Calculate the change in the position angle over the fifty-year period 2000 to 2050.

We have

α = 14h 39m 35.885s   (= 219°89952)
δ = –60° 50' 07".44   (= –60°8354)
μ_α = –0.49826 s   (= –0°00207608)
θ = 222°3
t_0 = 2000
t = 2050.

Carrying out the calculations gives $\Delta\theta$ = –0°.3 and hence θ_0 = 222°0. So there is a change of –0°.3 over the 50-year period due simply to procession and proper motion. (Note: 222°0 is not the position angle for the year 2050, but the position angle for the year 2000 referred to the pole of the year 2050.)

Differential Atmospheric Refraction

A further correction to both the position angle and the separation must be made, this time for the effects of atmospheric refraction. The correction should be made in reducing observations as well as when comparing observed with calculated values. The effects are negligible for small separations, as both components are subject to the same degree of refraction, and for stars of small zenith distance, where there is little displacement of star positions due to refraction.

The zenith distance of the star is given by

$$\cos z = \sin \delta \sin \phi - \cos \delta \cos \phi \cos H$$

where z is the zenith distance
ϕ is the latitude of the observer
H is the hour angle of the star
(= local apparent sidereal time – α).

The position of the star is measured towards the pole but it is displaced towards the zenith, hence the angular difference between these two directions, the parallactic angle, needs to be calculated:

$$\tan q = \frac{\sin H}{\tan \phi \cos \delta - \sin \delta \cos H}$$

where q is the parallactic angle.

The refractive index of the atmosphere varies according to temperature and air pressure.

Let C be the air temperature in degrees Celsius

    P be the air pressure in millimetres of mercury

    R be the refractive index.

Then

$$T = C / (C + 273)$$
$$A = 0.024 + 0.079017P - 0.0826PT$$
$$B = 0.004 - 0.0001101P - 0.000028PT$$
$$R = A + B.$$

The changes in the position angle and the separation can be found using Chauvenet's equations. These equations hold only for zenith distances less then 75°, i.e. for stars more than 15° above the horizon.

$$\Delta\theta = -R (\tan^2 z \cos (\theta - q) \sin (\theta - q) + \tan z \sin q \tan \delta)$$
$$\Delta\rho = \rho R (1 + \tan^2 z \cos^2 (\theta - q)).$$

Estimating Double Star Magnitudes

It is useful to provide estimates of the magnitudes of the components as well as the position angle and separation when measuring double stars. The magnitudes should be estimated to a tenth of a magnitude. A method for estimating the magnitudes is described in the *Webb Society Deep-Sky Observer's Handbook, Volume 1, Double Stars* (second edition), page 24.

The method is as follows: estimate the difference in magnitude between the two components, then with a low-power eyepiece, so that the double star appears as a single star, estimate the magnitude of the apparently single star. This will give the combined magnitude of the pair. The combined magnitude can be estimated by comparing the star with two other stars of known magnitudes in the field of view, in very much the same way that variable star observers make visual estimates of star magnitudes. (Such a method is described in *The Webb Society Deep-Sky Observer's Handbook, Volume 8, Variable Stars*, Chapter 3)

From the combined magnitude and the difference in magnitude the individual magnitudes can be determined. Let A and B be the magnitudes of the brighter and fainter components respectively. Let C be the combined magnitude and d be the difference in magnitude, i.e. $C = A + B$ and $d = B - A$. The magnitudes for the individual components can be found from

$$A = C + x$$
$$B = A + d.$$

A table for the different values of x for different values of d is given in the *Webb Society Handbook*. The values for x come from the formula

$$x = 2.5 \log_{10} (10^{-0.4d} + 1)$$

where d is the magnitude difference $B - A$, not $A - B$, i.e. $d > 0$.

The equatorial double 70 Ophiuchi appears as a single star of magnitude 3.8. When resolved through a telescope the components are found to have a magnitude difference of 1.8. The individual magnitudes are then found to be: for the primary, magnitude 4.0 and for the companion, magnitude 5.8

Providing magnitude estimates enables the stars to be monitored for any variation in brightness. Eta Geminorum and Alpha Herculis, for example, are visual binaries which each has a variable component.

Triple Stars

There might be occasions when triple stars are observed. Unfortunately the components are not always spaced sufficiently to measure from a single position. It is not always possible to measure the position of B with respect to A and then rotate the micrometer around and measure the position of C with respect to A. This is because multiple star systems tend to preserve their binary nature. If there are three stars then two of them form a binary while the third component usually orbits the other two as though they were a single star. Likewise, if there were a forth component it would normally be paired with the third component making a binary system where each component was itself a binary.

Measuring a triple star, then, usually entails measure B with respect to A and then measuring C with respect to the pair AB, or more specifically, with respect to the

centre of AB. The observation is made this way because
when sufficient magnification is used to separate A and
B the field of view is usually too small to include C and
conversely when the field of view contains C, A and B
are usually too close to be separated.

The problem, then, is to find the position angle of C
with respect to A. Fortunately the calculations involve
nothing more than some simple plane geometry.

Let $\Delta\theta = \theta_C - \theta_B$ where θ_C is the position angle of C
with respect to the mid-point of AB and θ_B is the posi-
tion angle of B with respect to A. Then θ_B

$$\eta = \rho_B\rho_C \cos \Delta\theta$$

where the subscripts B and C are as for the position
angles.

The separation of C from A is

$$\rho_{Ac} = \sqrt{\tfrac{1}{4}\rho_B^2 + \rho_C^2 + \eta}$$

and the position angle of C with respect to A is found
from

$$\tan \theta_0 = \frac{\rho_B\rho_C \sin \Delta\theta}{\tfrac{1}{2}\rho_B^2 + \eta}$$

$$\theta_{AC} = \theta_B + \theta_0.$$

Measurements of zeta Cancri for 2001 are

AB $\theta = 78°3$ and $\rho = 0''.86$ (2001.205, 8 nights)
$\tfrac{1}{2}$AB–C $\theta = 72°9$ and $\rho = 5''.79$ (2001.250, 7 nights)

($\tfrac{1}{2}$AB means the mid-point of A and B). Performing the
calculations gives

$$\Delta\theta = -5°4$$
$$\eta = 4.9573$$
$$\rho_{AC} = 6''.22$$
$$\theta_{AC} = 73°3.$$

The position angles, in this case, are all close to the
same value, which suggests that the three components
lie close to a straight line.

A Suitable Focal Ratio for Double Star Observing

Among the matters a double star observer must consider is that of a suitable focal length, or more

specifically a suitable focal ratio. When observing double stars it is important to have the image scale large enough to enable accurate measurements to be made. In other words, the two stars must be far enough apart at the focal plane to be seen readily and measured accurately. Unfortunately, when using a filar micrometer a small image scale cannot be compensated by an eyepiece giving a high magnification. The reason is that in magnifying the image the wires of the micrometer are also magnified. The point of having a large image scale is to keep the wires of the micrometer small with respect to the distance between the two stars. Imagine having an image scale so small that the separation of the two stars at the focal plane was less then the diameter of the micrometer wire – the wire could hide both stars at once. So what is required is an image scale large enough to make the separation of the two stars greater than the diameter of the micrometer wires.

Deriving a Value for the Focal Ratio

The image scale, s, is just the reciprocal of the focal length of the telescope, F, in radians. To convert radians to seconds of arc multiply by $648,000/\pi$. Hence the image scale in seconds of arc is

$$s = \frac{648,000}{\pi F}.$$

The desired image scale is one that will make the distance between two stars at the focal plane greater than the diameter of the micrometer wires. Well, what is the minimum separation between two stars? The answer depends on the resolution of the telescope.

The resolution of the telescope in seconds of arc is given by

$$r = \frac{120}{D}$$

where D is the aperture of the telescope in millimetres.

Now, let the image scale equal the resolution, i.e. equate the right hand sides of these two equations, and then multiply both sides by $F/120$. The result is

$$\frac{F}{D} = \frac{5400}{\pi} = 1718.87.$$

As F/D is the focal ratio of the telescope this equation gives the focal ratio at which the image scale equals the resolution. That is to say, this is the focal ratio at which the image scale is the resolution of the telescope per millimetre. So, for a 20-cm telescope, which has a resolution of 0.6″, a focal ratio of 1718.87 would give an image scale of 0.6 seconds per millimetre. However, unless the wires of the micrometer are 1 mm in diameter, a focal ratio this large will not be required. The required focal ratio will be 1718.87 multiplied by the diameter of the micrometer wires, i.e.

$$\frac{F}{D} = 1718.87w$$

where w is the diameter of the micrometer wires.

In the case of the RETEL micrometer, $w = 0.012$ mm. Hence, $F/D = 20.6$. This means that a focal ratio greater than 20.6 will ensure that the distance between two stars that the telescope can resolve will be greater than the diameter of micrometer wires.

There is a school of thought that holds that as the separation is measured with two wires, the separation at the focal plane should be at least twice the diameter of the micrometer wires. In the above case, this would mean that a focal ratio of 41.2 would be required.

Either way, a very long focal ratio is required. Telescopes do not usually come with focal ratios of this order. The way to effectively increase the focal ratio is to use a Barlow lens. The amplification factor of a Barlow lens is about 2–3 times. To achieve an effective focal ratio of 41.2 with a 3× Barlow would require a focal ratio of about 13.7, whereas an f/10 telescope with a 2× Barlow would achieve an effective focal ratio of 20.

These figures suggest that a telescope with a focal ratio of at least 10 would be required for double star observing, providing it is used with a Barlow lens of at least 2× amplification. If a telescope of a shorter focal ratio is used then the resolution of the telescope, for measuring double stars, is going to be limited by the size of the micrometer wires rather than by the aperture of the telescope.

Of course, if a double image micrometer were being used instead of a filar micrometer, then a shorter focal ratio, higher power eyepiece combination would be feasible. The focal ratio then would be limited by the focal length of the eyepiece being used. The focal ratio should be at least numerically equal to the focal length of the eyepiece in millimetres in order to achieve twice the resolving magnification.

So, if the eyepiece has a focal length of 9 mm the telescope should have a focal ratio of at least f/9. Conversely, if the telescope has a focal ratio of f/9 the focal length of the micrometer eyepiece should be 9 mm at the most. (This relationship between the focal length of the eyepiece and the focal ratio of the telescope holds also when using a filar micrometer. The size of the wires of the filar micrometer, however, dictate focal ratios that are numerically well in excess of the eyepiece focal length.)

There are, of course, double star observers whose instruments do not meet the above criterion, however it is something that a prospective double star observer should consider when deciding on what instruments to choose.

Observing Double Stars with an Alt-azimuth Mounted Telescope

The application of computer technology to telescope drives has enabled sidereal tracking to be automated on alt-azimuth mounted telescopes. Alt-azimuth mounted telescopes, however, turn about an axis through the zenith instead of an axis through the pole, as do equatorially mounted telescopes. This means that the fixed point on the celestial sphere for such telescopes is the zenith, instead of the pole. As a consequence of this stars in the field of the eyepiece rotate around the centre of the field as the telescope follows the stars across the sky. In the case of a double star this will cause the companion to circle the primary star in the course of the night.

An example of this field rotation, as it is called, is the belt of Orion. In northern latitudes, the three stars that form the belt stand vertically when the constellation is rising, but lie along the horizon when it is setting. In the southern hemisphere the orientation is reversed: lying when rising, standing when setting.

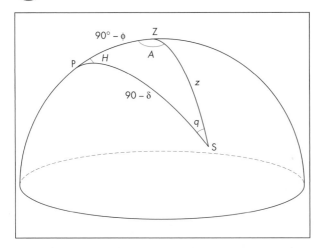

Figure 22.1. The Astronomical Triangle is formed by three points on the celestial sphere: the Pole (P), the zenith (Z), and a star (S). The sides of the triangle are PZ = 90° − ϕ (the co-latitude), PS = 90° − δ (the co-declination), and ZS = z (the zenith distance). The internal angles ZPS = H is the hour angle of the star, PZS = A is the azimuth of the star and PSZ = q is the parallactic angle.

The Parallactic Angle

In order to understand the problem we need to know something about the astronomical triangle. The astronomical triangle is formed by three points on the celestial sphere: the north celestial pole, the zenith and the star being observed. The angle that is of particular interest to us here is the angle subtended at the star between the pole and the zenith, i.e. the angle pole-star-zenith. This angle is known as the parallactic angle and is usually designated by the letter q. The parallactic angle increases as the hour angle increases. When a star is on the meridian $q = 0$ if it is on the equatorial side of the zenith, but $q = 180°$ if it is on the polar side. The reverse is the case for an observer in the southern hemisphere.

The parallactic angle of a star changes in the course of the night, due to its diurnal motion. Its value at any time, i.e. for any hour angle of the star, is given above in the section on calculating the effects of atmospheric refraction.

The Position Angle

As the zenith is the fixed point on the celestial sphere for an alt-azimuth mounted telescope, position angles measurements made with such a telescope would be referred to the zenith. Let us call position angle measurements made with respect to the zenith, then, the zenithal position angle to distinguish it from the position angle made with respect to the pole.

The direction of the zenith in the field of view can be determined by the same method that would be used to determine the direction of the pole with an equatorially mounted telescope, i.e. a star near the celestial equator is allowed to drift across the field of view, except in this case it must also be close to the meridian.

The position angle can be found by measuring it in the usual way, except, of course, that it is being measured with respect to the zenith. All that needs to be done in addition is to note the time of the observation, so that the parallactic angle can be determined and then subtract the parallactic angle from this zenithal position angle to obtain the position angle with respect to the north pole, i.e.

$$\theta = \theta_z - q$$

where θ is the position angle, θ_z is the zenithal position angle and q is the parallactic angle.

Field Rotation

The continual changing of the parallactic angle is known as field rotation and it is the main difficulty in measuring double stars with alt-azimuth mounted telescopes. The difficulty lies not so much in the fact that the orientation of the field is continually changing, but in the rate at which it is changing. The rate of field rotation, therefore, needs to be evaluated to determine the feasibility of being able to measuring the position angle accurately

The rate at which the parallactic angle is changing, i.e. the instantaneous rate of field rotation, can be found by differentiating the above equation for the parallactic angle. Hence,

$$\frac{\mathrm{d}q}{\mathrm{d}H} = \frac{15\cos^2 q(\tan\phi\cos\delta\cos H - E\sin\delta)}{(\tan\phi\cos\delta - \sin\delta\cos H)^2}$$
$$= \frac{15(\tan\phi\cos H - \sin\delta)}{\sin^2 H + (\tan\phi\cos\delta - \sin\delta\cos H)^2}$$

The constant, 15, converts the rate to degrees per hour. The second form of the equation enables the rate of field rotation to be found without having to find the parallactic angle.

Evaluating the derivative we find that the rate of field rotation peaks when the star crosses the meridian, i.e. when $H = 0$. Furthermore, the higher the star's culmination, i.e. the smaller the difference between δ and ϕ, the greater will be its rate of field rotation when it crosses the meridian. The maximum rate of field rotation, therefore, occurs when a star passes through the zenith. This implies that the worst time to observe a double is when it is best placed for observing! Consequently, there is a spherical cap around the zenith in which the rates of field rotation are too great to enable accurate measurements to be made. Field rotation rates close to the zenith can reach hundreds of degrees per hour. However, such high rates can only be sustained for very short periods (as they clearly cannot rotate more than 360° in 24 hours) after which they reduce to low rates again.

Conversely, the rate of field rotation is zero when the star crosses the prime vertical, i.e. when the star is due east and again when it is due west. Obviously, only stars with declinations that lie between the observer's latitude and the celestial equator will cross the prime vertical. Hence, the best times to observe double stars, as far as field rotation rates are concerned, are when the stars are in the eastern and western regions of the sky.

The average rate of field rotation is, not too surprisingly, 15° per hour. This is half the rate of 30° per hour at which the hour hand of a clock turns. A rotation rate of 360° would be a very high rate, yet it is the rate at which the minute hand of a clock turns.

The problem, then, lies not in whether the rotation rate is too great to make a position angle measurement, but in whether the observation can be timed with sufficient accuracy, i.e. in recording the time when the companion was at that particular zenithal positional angle. For a star with a field rotation rate of 15° per hour, the time of the zenithal position angle measurement would have to be made to an accuracy of 12 seconds; that is to say that the time will have to be noted within 12 seconds of having set the position angle on the micrometer if an accuracy of 0.1 arcminutes is to be achieved. This is because the position angle would have rotated 0.1 arcminutes in 24 seconds and after 12 seconds the position angle will be nearer the next tenth of a degree. In practice one would set the positional angle and then note the time before taking the positional angle reading.

The rate of field rotation that can be tolerated will depend upon how accurately the observation can be timed. If it is done manually and we assume that the time can be read off the clock within 10 seconds of making the position angle setting then we have an upper limit on the rate of field rotation of 18° per hour. Field rotation rates less than this are typically found in the eastern and western sections of the sky. If the time is recorded electronically then much higher rates can be tolerated and the "no go" area around the zenith could be reduced considerably.

The highest rate of field rotation, in degrees per hour, that can be tolerated is just 180° divided by the number of seconds it takes to note the time of the observation, or conversely, divide 180° by the field rotation rate to determine the time limit.

The Separation

The separation can be made in the usual way. However, to make the double distance measurement set the fixed wires on the primary with the position angle wire bisecting the primary and the companion. Then move the moveable wire onto the companion. Note the reading on the micrometer screw. Now rotate the micrometer right around so that the position angle wire again bisects the primary and the companion, but the moveable wire is on the opposite side of the primary to the companion. Then move the movable wire back, across the primary, to the companion again. Note the new reading on the micrometer screw. The difference between the two readings gives a measure of the double distance.

Ideally, the companion should be on the position angle wire when the separation measurement is made, but due to field rotation it might have moved away.

The error this would induce would depend on the separation. The error is just $\rho(1 - \cos \Delta q)$, where ρ is the separation and Δq is the change in the parallactic angle. If, in the time it took to move the moveable wire on to the companion, the companion had moved two degrees it would induce an error of $0''.004$ in a separation of $10''$. As two degrees represents four minutes at a field rotation rate of $30°$ per hour field rotation would not be a major source of errors in the separation.

Errors

Measurements of double stars made with an alt-azimuth mounted telescope are subject to the same errors as those made with an equatorially mounted one. However, additional errors can be introduced in converting the zenithal position angle to the position angle. The observer's latitude and the equatorial coordinates of the star are required to calculate the parallactic angle. The accuracy to which these are known determines the accuracy to which the parallactic angle can be calculated and in turn sets a limit on the accuracy of the position angle.

The errors in the parallactic angle would be negligible if the zenithal position angle was timed accurately and if the latitude could be determined accurately (perhaps from an accurate survey map or a GPS). Furthermore, precessing the right ascension and declination of the star from the catalogue positions would ensure accurate values for the coordinates of the star.

The separation, of course, will not be affected by these factors. Neither will the position angle if a mechanism that compensates for field rotation (a field de-rotator, as one manufacturer calls it) is fitted to the telescope. However such a compensating mechanism would, as it rotates, cause a right-angled eyepiece holder to "fall over", placing the eyepiece at an awkward angle. This would not be a problem if a right-angle eyepiece holder was not used, such as when viewing straight through the telescope or using a camera.

Computer Programs

The formulae presented here are implemented in a suite of computer programs that can be found on the

accompanying CD-ROM. Some additional programs
are included, such as calculating the visual aspect of a
double star (Chapter 7) and the calibration of the ring
and filar micrometers and the reduction of observations made with them (Chapters 12 and 15). The programs do not require any installation process, they can
be simply copied across to the computer hard disk
drive.

References

1 Green, Robin M., 1985, *Spherical Astronomy*, Cambridge
University Press, Cambridge, 282, problem 11.4.

Further Reading

Dating Observations

Astronomical Almanac Supplement, 1984, *Section S: The
Improved IAU System*, Washington: US Government
Printing Office and London: Her Majesty's Stationery
Office.

Precessing the Position Angle

Couteau, Paul, 1981, *Observing Visual Double Stars*, trans.
Alan Batten, MIT Press, Cambridge, MA, 121–2.

Differential Atmospheric Refraction

Argyle, R.W., 1990, *Webb Society Observing Section Reports*,
No. 6.

Estimating Double Star Magnitudes

Argyle, R.W., 1986, *Webb Society Deep-Sky Observer's
Handbook*, vol. 1: *Double Stars*, 2nd edn, ed. Kenneth Glyn
Jones, Enslow Publishers, Inc.
Greaney, M.P., April 1997, *Quarterly Journal of the Webb
Society*, 108.
Isles, John E., 1990, *Webb Society Deep-Sky Observer's
Handbook*, vol. 8: *Variable Stars*, ed. Kenneth Glyn Jones,
Enslow Publishers, Inc.

A Suitable Focal Ratio for Double Star Observing

Couteau, Paul, 1981, *Observing Visual Double Stars*, trans. Alan Batten, MIT Press, Cambridge, MA.

Observing Double Stars with an Alt-azimuth Mounted Telescope

Greaney, M.P., October 1993, *Quarterly Journal of the Webb Society*, No. 94.

Greaney, M.P., October 1995, *Quarterly Journal of the Webb Society*, No 102.

The star positions used in the examples were obtained from *The Bright Star Catalogue*, 5th rev. edn (Hoffleit +, 1991) on the Astronomical Data Center website http://adc.gsfc.nasa.gov/cgi-bin/viewer/specify.pl?file=catalog.dat&catalog=5050

Chapter 23

Star Atlases and Software

Owen Brazell

Introduction

As with all astronomical objects the challenge with observing double stars, except for the very brightest, is finding them. Luckily most commercially available star charts do plot the brightest double stars.

Paper Star Atlases

The cheaper paper star atlases tend to plot most double stars brighter than say magnitude 6.5 and they are traditionally marked as a circle with a line through it. This may or may not indicate the position angle of the star itself at its last measured epoch. There is usually no other labelling of the star to indicate what its designation is. The kinds of star atlas that fall into this league are the *Cambridge Bright Star Atlas* and *Norton's Star Atlas* although the latter does give information about double stars marked on the accompanying pages. The source material for the double star information is not often given. It is probable that in most cases the data was taken from a version of the WDS or its predecessors. The more advanced paper star charts such as *Sky Atlas 2000* and *Uranometria 2000* use the same database of objects; they culled all double stars with a total magnitude less than 8.5 from a 1976 version of the IDS catalogue. This may not be too much of a problem as most of the stars brighter than this limit that are likely to resolved by common amateur instruments will probably be in the list, although the positional and orbital data may now be suspect. The stars are marked on the charts, again with the standard symbol but with no labelling information on the star chart itself as to what the double star designation was. The information on this was contained in the accompanying reference guide *Sky Catalogue 2000 Volume 2* which listed all the double stars marked on *Sky Atlas 2000*. A similar procedure was used for the older *Atlas Coeli* with its accompanying catalogue. The *Sky Atlas 2000 Companion* released to accompany the new version of *Sky Atlas 2000* does not list any double stars. The Deep-Sky Field Guide released to accompany the first edition of *Uranometria 2000* also contains information only on non-stellar objects and nothing on double stars. Probably the last major paper star atlas, The *Millennium Star Atlas* plots double star data taken from the *Hipparcos* mission. Unlike previous paper atlases it treats double and multiple stars in a more complex way. Even though the stars are still not labelled each double star has a radial line whose length is logarithmically proportional to the separation of the components and whose angle represents the position angle. The data are taken from the separation and position angle measured by the *Hipparcos* satellite at epoch 1991.25. The positional data for double stars are also taken from the *Hipparcos* catalogue which makes for an improvement over positions taken from the original sources. Although there is supposed to be a new version of *Uranometria 2000* due out late in 2001 it is likely that with the *Millennium Star Atlas* the age of the large scale printed star atlas really came to an end and computer star charting programs took over, certainly for the more demanding users.

Computer Databases

Before discussing computer star charts it worth looking at the raw data itself. Most computer star charting programs use the raw data from the *Washington Double Star Catalogue*, or WDS for short. This database which is updated daily with new measures and recently discovered double stars is maintained by the United States naval Observatory and is available through the Internet at http://ad.usno.

navy.mil/wds/wds.html. This site includes the database itself which can be downloaded, although at the time of writing (October 2002) it was 11.5 Mb, along with information on the structure of the catalogue and a method of querying it. In order to help get information into a form useful for amateurs to study several people have written access programs to this data. Most of these databases are based on the 1996 release of the catalogue which was the last sent out on CD-ROM. Probably the best is a program put out by the hard-working group at the Saguaro Astronomy group in Arizona. They have generated a double star program similar to their better known SAC database for deep-sky objects called SAC DB 2.1 which was an attempt to provide a working list of double stars for owners of modest telescopes.

The database was based on a version of the WDS from 1991 and took double stars with primaries whose magnitudes were greater than magnitude 9 and secondaries whose magnitudes were greater than 13. They then wrote a small database program which allows the user to query the database based on constellation, magnitude or just by star designation and get information on that star. A typical screen from that program is shown in Figure 23.1. The program is available for free from their website at www.saguaroastro.org. It is still a DOS type utility. As double stars became of more interest several other utilities became available which allowed the user to either input data or take data from

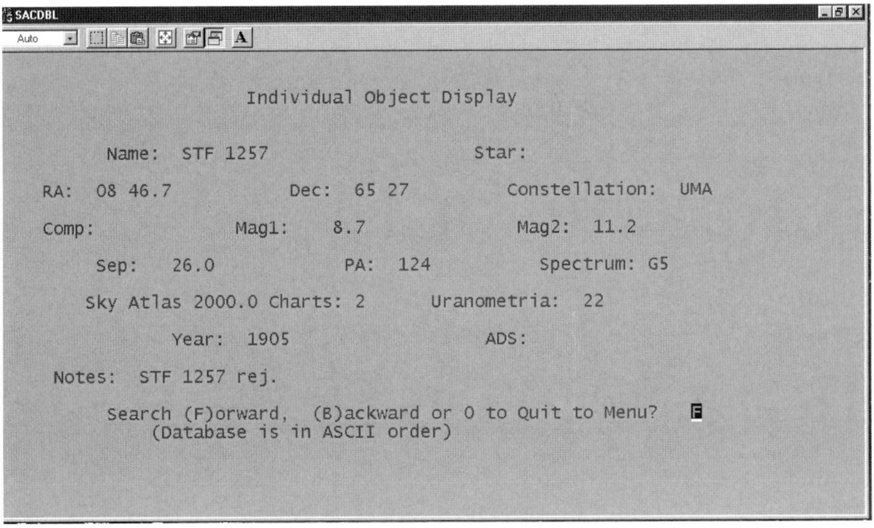

Figure 23.1. An example from the Saguaro database.

```
 SACDBL                                                        _ 8 X
 Auto

                         Individual Object Display

           Name:  STF 1257                  Star:

    RA:   08 46.7              Dec:  65 27        Constellation:  UMA

    Comp:              Mag1:    8.7              Mag2:   11.2

         Sep:   26.0              PA:  124            Spectrum: G5

         Sky Atlas 2000.0 Charts: 2        Uranometria:  22

              Year:  1905                    ADS:

    Notes:   STF 1257 rej.

         Search (F)orward,  (B)ackward or 0 to Quit to Menu?
              (Database is in ASCII order)
```

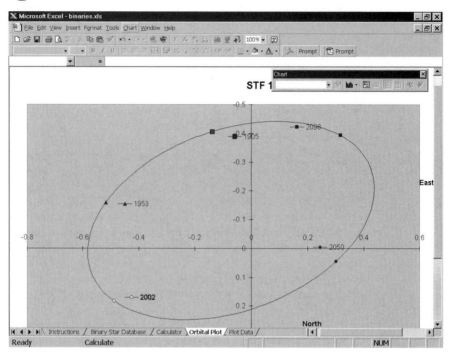

the orbital catalogues and generate positions or orbits based on current data.

Figure 23.2. An example of the software by Brian Workman.

The example shown in Figure 23.2 is the data taken from a useful Excel spreadsheet by Brian Workman available via http://www.psiaz.com/polakis/index.html. Another useful site is Richard Dibon-Smith's orbital pages at http://www.dibonsmith.com/orbits.htm. For raw data on double stars the data from the *Hipparcos* mission has been made available on a CD called *Celestia 2000* which is available from Sky Publishing in North America or from ESA for the rest of the world. This CD allows you to search on the *Hipparcos* catalogue for double star data and provide a rudimentary plot of this data. The plotting functionality is not as great as the mainstream programs but the data is some of the more accurate positional data for the brighter stars that *Hipparcos* measured.

Computer Star Charts

Computer star charting programs come in two different flavours: the multimedia directed ones such as

Redshift 4 and Starry Night and the observer orientated ones such as SkyMap Pro, Guide, Megastar and The Sky. Neither Redshift 4 nor Starry Night Deluxe appeared to have any double star information at all. Even if you click on a bright double star such as gamma Andromedae nothing comes up to indicate its multiple nature. These programs are more aimed at the educational market than the observers' market.

Most of the current crop of high-end computer star charting programs at the time of writing (Summer 2001) use as their base double star catalogue the 1996 version of the WDS. That is likely to change as newer releases of most of these programs are due out towards the end of 2001 and I would expect them to use the 2001 version of the WDS. The main differences between the programs are then how they allow the user to search for and display the data. Running through the four most popular star charting programs will give a flavour for how they display their data and how the search for that information proceeds. The same star, STF 1257, a double star from F.G.W. Struve's catalogue, has been chosen for all the displays.

We start with the oldest, in its current form, which is Megastar Version 4. Megastar is similar to Guide in that when you search for a double star it displays a list of discoverers and then you select the number. A typical screen is shown in Figure 23.3. The display of data is then shown in Figure 23.4. Megastar displays the least information about the star but does label the object with its correct name and has a symbol to show where the star is.

Guide Version 7.0 from Project Pluto does not appear to allow the star to have a double star symbol but it does allow the labelling of double stars with their names from the WDS. The example of STF 1257 is shown in Figure 23.5.

Unlike all the other programs Guide does allow the display of data on the chart from the *Catalogue of Components of Double and Multiple Stars* or CCDM. The main reason for using this catalogue is to get more accurate positions for the various components of the double stars listed in the CCDM. The CCDM is a rather strange catalogue in the information it contains and its use is mainly in improving the positional data. This was provided as an extra dataset with Guide 7. Guide also lists basic data from the CCDM. Where Guide scores over Megastar is in the information that it

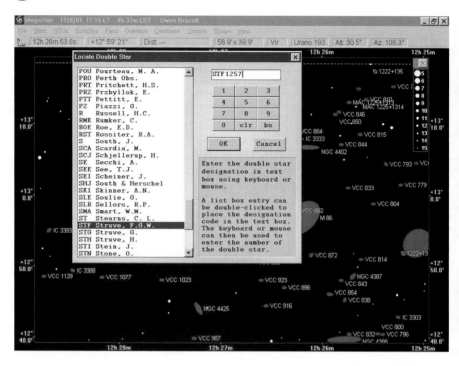

Figure 23.3. A typical screen display from Megastar.

provides about the double star itself and this is illustrated in Figure 23.6.

Guide 8 also now includes the WDS 2001 edition and uses the catalogue of double star orbits as well to give information on the current PA and separation of double stars as well as animations of the stars in time showing orbital behaviour. Guide 8 also provides a user data set with the double stars from the Astronomical leagues Double star list.

The popular SkyMap Pro program in Version 7 allows the user to search for a double star through the generic star designation search facility in which the discoverer's mnemonic needs to be known first. The program then shows the double star with the standard paper chart symbol of a circle with a line through it but does not appear to allow the star to be labelled with its double star designation. SkyMap Pro also displays a large amount of information about the star including data from the CCDM and the WDS in its standard star data window, as shown in Figure 23.7.

With the appearance of SkyMap Pro 8 late in 2001 several other features of interest to double star observers have been added. These include updating the main double star catalogue to the WDS 2001, the ability

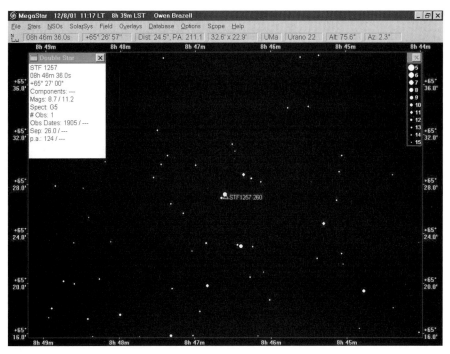

Figure 23.4. Data display in Megastar.

to label double stars with their real names rather than their *Tycho* numbers and, perhaps most interestingly, the inclusion of the fifth catalogue of double star orbits which allows the program to compute the positions and separations for 1500 double stars whose orbits are known accurately at the current time.

The most advanced of all the sky charting programs, The Sky, version 5, level IV, from Software Bisque, is also the most limited when it comes to double stars. There is no search facility for double stars by the WDS designation. They can be queried by another stellar catalogue name if they have one. The program should mark double stars with a special symbol but also appears to display little or no information about the stars themselves. For the double star observer this program could not be recommended. The Sky for PocketPC contains no double star information at all.

The new kid on the block in star charting/logging software is SkyTools. This software is perhaps different in that in its current form, version 1.5, it allows the user to search the catalogues if you know the name of a double star from another catalogue and it will display an impressive amount of information. A user downloadable file will give you the cross-reference from the

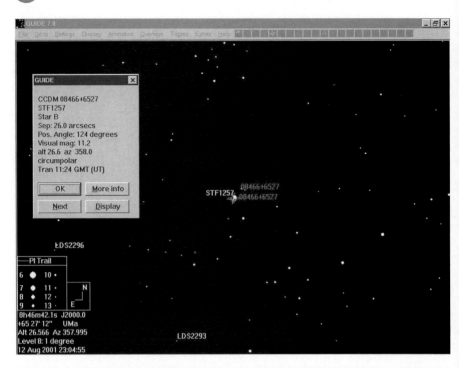

Figure 23.5 (above). An example from Guide Version 7.
Figure 23.6 (below). An example of data displayed in Guide Version 7.

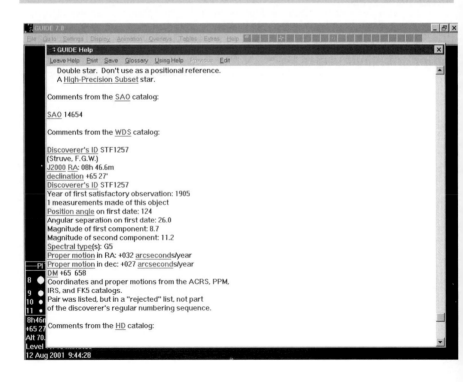

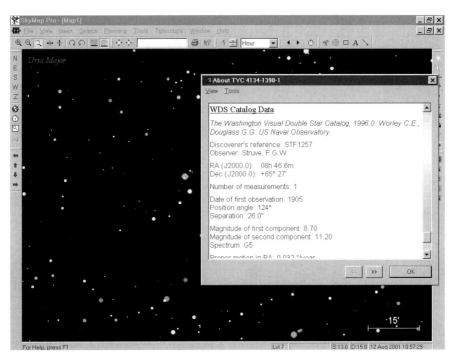

Figure 23.7. An example from SkyMap pro.

WDS to one of the catalogues that SkyTools will display, normally an HD number. The base double star catalogue in SkyTools is also different being, unusually, the CCDM supplemented with double star data from *Hipparcos*. SkyTools is also unusual in that it has many different charts that can be plotted but on none of them as far as I could see did the selected double star have a symbol to indicate that it was a double, nor did there seem to be a way of labelling the star with its ID.

Despite the fact that the information is available in the catalogues it is disappointing that none of the programs plot the double star symbol in the way that the *Millennium Star Atlas* does with a line indicating the Position angle and its length giving the separation at its last measured epoch. With the exception of The Sky all the programs do something different for the double star observer and a choice of which one to use may come down to the other facilities offered.

Uranometria 2000 (2nd edition) has come out since the main body of the text was written and its double star data was taken from the *Hipparcos* and *Tycho* catalogues using PAs and separations from those catalogues along with hand updates for pairs wider than 60″.

Star Atlases and Software

SkyMap Pro (www.skymap.com/)
The Sky (www.bisque.com)
Guide (www.projectpluto.com)
Megastar (www.willbell.com/software/megastar/index.htm)
SkyTools (www.skyhound.com)
ECU (www.nova-astro.com)
Starry Night (www.siennasoft.com)

Chapter 24

Catalogues

Bob Argyle

Northern Hemisphere

The first catalogue of double stars is due to Christian Mayer in 1779 and contains 80 entries. It was the work of Herschel and especially Struve who gave the whole subject a respectability which was lacking. Struve's *Mensurae Micrometricae* (to give the catalogue its shortened name), which appeared in 1837, was a huge work in more than one respect (Figure 24.1).

The next major catalogue did not come until 1906 when Sherburne Wesley Burnham produced his *A General Catalogue of Double Stars within 121 Degrees of the North Pole*, published by The Carnegie Institute of Washington. It contains 13,665 systems and is unique in that it includes all known references to the measures contained within. It did, however, include some wide pairs which were not binary but optical in nature.

In 1932, Robert Grant Aitken produced the *New General Catalogue of Double Stars within 120 Degrees of the North Pole* with 17,180 entries. It is usually known as the ADS. The limits for inclusion were stricter than those of Burnham so Aitken's catalogue contains more true binary systems.

Southern Hemisphere

In 1899 at the Royal Observatory at the Cape of Good Hope, R.T.A. Innes published A *Reference Catalogue of*

Southern Double Stars, but chose rather narrow limits to decide which pairs went in. In 1903 Innes became Government Astronomer at the Union Observatory Johannesburg and in 1925, ably assisted by W.H. van den Bos and W.S. Finsen, started on a new survey for double stars in the southern skies using the newly installed 26.5-inch (67-cm) Grubb refractor. Innes compiled the *Southern Double Star Catalogue* in 1927 as a means of identifying new double stars during the subsequent searches. This covered the zones −90° to −19° and contained 7041 systems.

In 1910 R.P. Lamont a wealthy industrialist and friend of the double star observer W.J. Hussey (who was latterly Director of the Observatory of the University of Michigan) had authorised plans for a large telescope for double star observation. Hussey planned to install it at Bloemfontein in South Africa to continue his own searches for new double stars. Tragically Hussey died in 1926 en route to South Africa but the project was taken over by R.A. Rossiter who stayed until 1952. Rossiter then compiled the *Catalogue of Southern Double Stars*, essentially a list of the pairs discovered by Rossiter and his assistants Donner and Jessup (Figure 24.2) – more than 7600 in the 24 years ending 1952.

Figure 24.1. Julie Nicholas, formerly Librarian at the Institute of Astronomy, with copies of Struves' first catalogue, the IDS (open on the desk), and the WDS (on CD-ROM). The latter could also contain every measure ever made. (R.Sword (IOA))

Figure 24.2. These 5 observers were responsible for more than 10,000 double star discoveries. Pictured outside the Lamont–Hussey Observatory in September 1928 are (left to right) H.F.Donner, W.S.Finsen, R.A.Rossiter, W.H.van den Bos and M.K.Jessup.

All-Sky Catalogues

The first all-sky catalogue of double stars did not appear until 1961. It is printed in two volumes as Volume 21 of the *Publications of Lick Observatory* and its formal title is *Index Catalogue of Visual Double Stars 1961.0*. It is still the only printed version of an all encompassing catalogue and is now likely to remain so given that it runs to 1400 pages of closely printed script. Edited by Hamilton Jeffers, Willem van den Bos and Frances Greeby, the *Index Catalogue of Double Stars* or IDS was issued to include the large number of discoveries that had been made at the Republic and Lamont–Hussey Observatories in South Africa (Figure 24.3).

With the development of the *Hipparcos* project in the 1970s it was apparent that with the very approximate positions (0.1 minutes of time in RA and 1–2′ in Declination) and insufficient cross references between the IDS and other catalogues – largely the *Durchmusterungs* – it would be a disadvantage when programming the satellite to observe double and multiple systems. This led Jean Dommanget, a member of the INCS (Input Catalogue) consortium and a well-known double star researcher at the Royal Observatory in Brussels to propose a new catalogue – the CCDM (*Catalogue of the Components of Double and Multiple Stars)* which would feature considerably better

Figure 24.3. Dr. W.H. van den Bos looks on proudly as the President of the South African Council for Scientific and Industrial Research, Dr. S. Meiring Naudé, peruses a copy of the IDS (1968) (copyright CSIR).

positions and photometry for the stars in the *Hipparcos* input catalogue (about 120,000) which were known to be double or multiple. More importantly it was necessary to list all the components of each system so that the new discoveries made by *Hipparcos* could be evaluated more easily. The purpose of the CCDM is to be complementary to the WDS. It does not aim to be all-inclusive but it does contain more detailed information on a smaller number of systems. In collaboration with Omer Nys, Jean Dommanget produced the first version of CCDM in 1994 and a second version appeared in 2002 which contains 49,325 systems.

The current version of CCDM can be found via the CDS at Strasbourg at

http://cdsweb.u-strasbg.fr/cgi-bin/Cat?I/269A and a file of all the systems observed by the *Hipparcos* satellite which is essentially a subset of the CCDM can be found at http://cdsweb.u-strasbg.fr/cgi-bin/Cat?I/260

The central data repository for visual double star data continues to be kept at the United States Naval Observatory. In the early 1960s the late Charles Worley received the Index Catalogue (in card form) from Lick Observatory which had appeared in two parts as described above. Copies of this catalogue were rarely seen except in the reference libraries of observatories so data on visual double stars was not easy to obtain at this time.

Worley (Figure 24.4), ably assisted by Geoffrey G. Douglass and others, spent the rest of his working career bringing the Lick Catalogue up-to-date. This

Figure 24.4. Charles Worley, pictured at Santiago de Compostela, in July 1996. (Angela Argyle)

meant, amongst other tasks, converting the punch cards into computer files, inputting new measures and discoveries on a regular basis and weeding out errors. The result of this was the first electronic version of the *Washington Double Star Catalogue*, WDS 1996.0 – so called because it represented the state of the data archive at the beginning of 1996. It had grown to some 78,000 entries so producing a printed copy was out of the question. After Worley's death the archive was taken over by Dr. Brian D. Mason who had done his research in the discipline of speckle interferometry at Georgia State University. Dr. Mason and his team have recently produced the WDS 2001.0 and are issuing incremental updates at regular intervals. The current catalogue contains the new pairs discovered by the *Hipparcos* satellite and so offers double star observers a whole new set of pairs to measure. Most of these pairs have remained unobserved from the ground but it must be noted that many are very difficult and require both large apertures and good seeing.

The USNO have recently produced a CD-ROM which contains, amongst other useful files, the WDS 2001.0 catalogue. This is the version of the WDS frozen as of late Sept 2001 and contains a single-line entry for each of 84,486 systems. The updated version of the WDS catalogue can be downloaded from the USNO site at http://ad.usno.navy.mil/ad/wds/wds.html and is currently 11.5 MByte in size (98,084 systems). It is also available in portions of 6 hour intervals in RA. Another file contains useful notes on systems of interest. In addition there are two useful cross-reference files, for *Hipparcos* v HDS numbers (*Hipparcos Double Stars*) and WDS v ADS. Although the latter reference number is not adopted in the current WDS, it is still used by orbit-computers. The data is given in a rather compact form and so on first acquaintance it needs the use of the accompanying key to decipher what the data columns mean. The main advantage of the online WDS is that it is not only obtainable by anyone (try finding a copy of the IDS!) but it is a dynamic database and is updated regularly!

For those not on the Internet then the enclosed CD-ROM contains a recent edition of the on-line WDS catalogue together with the *Sixth Orbit Catalogue* and the *Fourth Interferometric Catalogue*.

Whilst the WDS catalogue is large, it is dwarfed by the *Observations Catalogue* which is also maintained by USNO but which is not generally accessible. At the time of writing (September 2002) this consisted of 585,261 mean observations of 98,084 pairs. Requests for data can be made using the request form on the website. This is particularly useful for orbit determinations for instance.

Interferometric Data

The *Third Catalog of Interferometric Measurements of Binary Stars* is made up of all observations made by interferometric techniques whether it be speckle, ground-based arrays or even the early Michelsen Interferometer observations at Mount Wilson. It also includes data from the *Hipparcos* and *Tycho* catalogues. The common property is that the accuracy is extremely high and this is an ideal source of useful data for those who want to test the quality of their telescopes. The author has selected several hundred pairs from this list which show very little motion and thus can be used as a resolution test. The separations range from 0.2 to 2″. A subset can be found in Chapter 2.

Perhaps a more useful set of measures for those with a small telescope is those made by the USNO Astrometry Department using a speckle interferometer on the 26.5-inch refractor at Washington. Since 1990 and again under the direction of Charles Worley, an extensive programme of measurement of brighter binaries has been undertaken and the results have appeared in a series of papers in the *Astronomical Journal and Astrophysical Journal Supplements* (see, for example, Mason et al.[1]).

Double Star Nomenclature

Many observing guides tend to use the old catalogue names for double stars, some of which use Greek letters, i.e. β for Burnham, φ for Finsen and so on. This has nothing to do with the Flamsteed letters such as δ Equulei but the current nomenclature in the Washington Double Star catalogue avoids such possible complications and tends to be favoured by the professional observers. In this scheme the star is referred to by its J2000 coordinates. Thus as an example we can take Castor which is Σ1110 (Σ being the Struve catalogue) but appears in the WDS catalogue as STF1110 where the discoverer is denoted by one, two or three letters and avoiding Greek names altogether. The WDS name for Castor is thus WDS07346+3153AB.

At the time of writing a vigorous debate is taking place in which a nomenclature that will account not only for visual double stars but spectroscopic and other kinds of pairs and exoplanets has been proposed. A scheme based on a modified WDS is being prepared for discussion at the next IAU in 2003.

In many cases the ADS (or *Aitken Double Star Catalogue* number) is still used but this system is no longer supported by the WDS, partly because it includes only 20% of all known pairs. References to the *Burnham Double Star Catalogue* (BDS) numbers are also occasionally to be found but again not in the WDS.

The short list below can be used to identify the abbreviated catalogue names used in this book in earlier chapters but it is not a complete list; this can be found on the WDS web page. The WDS system

Table 24.1.

Discoverer	Usually	WDS
Aitken, R.G.	A	A
Bos, W.H. van den	B	B
Brisbane Observatory	Brs0	BSO
Burnham, S.W.	β	BU
CHARA	CHR	CHR
Dunlop, J.	Δ	DUN
Herschel, W.	H I, II etc	H 1,2 etc
Hough, G.W.	Ho	HO
Howe, H.A.	Hwe	HWE
Hussey, W.J.	Hu	HU
Kuiper, G.P.	Kui	KUI
Krueger, A.	Kr	KR
Lacaille, N	Lac	LCL
Luyten, W.	LDS	LDS
McAlister, H.A.	McA	MCA
Piazzi	Pz	PZ
Rossiter, R.A.	Rst	RST
South J. and Herschel, J.	Sh	SHJ
Struve, F, G.W.	Σ	STF
Struve Appendix Catalogue I	Σ I	STFA
Struve Appendix Catalogue II	Σ II	STFB
Struve, Otto	OΣ	STT
Pulkova Appendix Catalogue	OΣΣ	STTA

usually consists of three letters and 4 numbers so if, for example, you wish to search the WDS for Dawes 4 then you need to look for the string DA - - - - 4 where the four blanks (-) are significant. In order to deal with the numerous designations of Herschel double stars and the various Pulkova catalogues, recent modifications have been made and these are shown in Table 24.1. In any case the ID string can be read as character format A7.

Greek Alphabet

α	alpha	β	beta	γ	gamma	δ	delta	ε	epsilon	ζ	zeta
η	eta	θ	theta	ι	iota	κ	kappa	λ	lambda	μ	mu
ν	nu	ξ	xi	o	omicron	π	pi	ρ	rho	σ	sigma
τ	tau	υ	upsilon	φ	phi	χ	chi	ψ	psi	ω	omega

References

1   Mason, B.D., *et al.*, 2001, *Astron. J.* **122**, 1586.

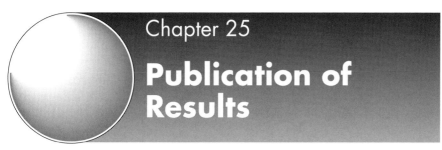

Chapter 25

Publication of Results

Bob Argyle

Introduction

Publishing observations of double stars is a natural consequence when an observer feels confident enough in the quality of his or her measures that they feel it is time to share them with the rest of the astronomical community. A lot of effort has gone into this work so it is only fair that the observer should gain credit for it. There is no fixed formula which can be applied to decide whether measures are of publishable quality or not. But recent lists of bright, close (0.5 to 2″) pairs (Mason *et al.*[1]) are available, so some comparison can be made to check on how good the agreement is. Other factors to consider include whether a particular pair has been observed many times or virtually ignored since discovery. A really accurate measure of a bright, relatively fixed, over-observed pair will not be as useful as a less accurate measure of a pair which has been ignored for 100 years or more, especially if it turns out that the latter has significant motion.

Measures can be published in several formats and in both professional and amateur journals but one thing cannot be over-emphasized. It is absolutely vital that the same measures are *never published more than once* since it can cause great confusion to the astronomers who collate all measures of visual binary data for the *Observations Catalogue* at the USNO in Washington. An example showing the publication of double star observations by the Webb Society can be seen in Figure 25.1. These data are the raw measurements and

Star	RA	Dec	Mags	PA	Sep	Epoch	N	Obs
Σ2434A-BC	1902.7	−0043	8.4, 8.8	92.8	26.40	2000.44	3	RA
Σ2455AB	1906.9	+2210	7.2, 9.4	28.7	8.42	2000.64	4	RA
β248AB	1917.7	+2302	5.5, 9.0	131.3	1.74	2000.68	4	RA
Σ2497	1920.0	+0535	7.6, 8.5	356.3	30.3	2000.62	4	JC
Σ2498AB	1920.2	+0403	8.4, 9.0	65.5	12.15	2000.68	4	JC
H84	1939.4	+1634	6.4, 9.5	300.7	28.43	2000.68	3	RA
Σ2585AB-C	1949.0	+1909	5.0, 9.0	310.0	8.05	2000.68	4	RA
Σ2596	1954.0	+1518	7.0, 8.3	299.2	2.00	2000.10	4	RA
Σ2607AB-C	1957.9	+4216	6.4, 8.9	289.9	3.1	2000.63	4	JC
S730AB	2000.1	+1737	7.0, 8.4	14.0	112.45	2000.69	3	RA
S730AC	2000.1	+1737	7.2, 9.5	337.1	78.12	2000.72	3	RA
H100AB	2000.2	+1730	9.5,10.3	256.1	24.11	2000.78	4	JC
H100AC	2000.2	+1730	9.5, 5.6	295.7	113.7	2000.81	4	JC
Σ2615	2002.9	+0824	7.9,10.8	304.3	9.15	2000.64	4	JC
Σ2653	2013.7	+2414	6.6, 9.5	275.4	2.7	2000.68	4	JC

Figure 25.1. An example of double star measures published in the Webb Society *Double Star Section Circulars.*

other tables contain notes on systems of interest and residuals from known orbits where applicable.

The paper should contain details of the micrometer type, the instrumental constant and the magnification employed.

The format of any list should contain the following information:

Identifier	Currently the standard is the WDS format (see Chapter 24). This also includes the J2000 position.
Catalogue	An alternative identification, not always necessary but it can be useful when using star atlases and catalogues such as *Burnham, Webb's Celestial Objects* and the *Sky Catalogue 2000.0.*
Mean position angle	This should be the mean value from the individual nightly values. Usually quoted to one decimal place for visual work but CCD astrometry may justify more. Avoid using angles greater than 360.0.
Number of PA measures	The number of independent nights from which the mean is formed. This will usually be the same as the number of nights used for the mean separation.
Mean separation	In arcseconds, usually quoted to two decimal places – if the observer con-

siders this to be a fair reflection of the scatter in the individual measures.

Number of separation measures	As for position angle. It may be for a highly inclined binary where the change is nearly all in separation that more measures in separation would be a sensible approach. Usually quoted to two decimal places in visual work.
Mean epoch	This is much easier to work out if each individual night is converted to a decimal of a year in the observing log. A day is 0.0027 of a year so midnight on 2001 Jan 10, for example, is 2001.027. It is quite sufficient to use the midnight value for that night and in fact mean epochs can be quoted to two decimal places for most small telescope observations. Chapter 22 contains a short algorithm to calculate decimal date from calendar date – use program JD&Epoch from the CD-ROM.
Observer	This will usually be given at the head of the paper for a single author. Usually a two-letter code is inserted at the end of each line in the data table if the list contains the measures of more than one observer. Those whose measures are included in the WDS *Observations Catalogue* are given a three-letter identifier by the compilers.
Orbit residuals	The differences (observed – computed) for both position angle and separation from the orbit for every epoch of observation. Include the author of the orbit and its date of publication. Include other orbits if there is little to choose between them. The CD-ROM contains the latest version of the orbital elements catalogue published by the USNO.

The following data can be given depending on taste:

(i) Difference in magnitude: usually estimated visually to 0.1 magnitude.

(ii) Standard error of position angle and separation. calculated from the individual measures that make up the means.

(iii) A note of whether the eyes were vertical to the wires (:) or parallel to the wires (..) when the observations were made.

(iv) The quality of the night – transparency and seeing, for instance.

References

1   Mason, B.D., et al., 2001, *Astron. J.* **122**, 1586.

Appendix

Some Useful URL Addresses

Washington Double Star Catalogue – USNO (http://ad. usno.navy.mil/wds/wds.html)

Primary source of data on any double star. Suggested lists of neglected and unconfirmed pairs for observing. Current version of the orbital catalogue (Sixth) is at http:// ad.usno.navy.mil/wds. Third speckle catalogue CD-ROM (June 2001) can be obtained by e-mail or post from: Dr. Brian D. Mason , Project Manager, Washington Double Star Program, Astrometry Department, US Naval Observatory, 340 Massachusetts Avenue NW, Washington DC 20392-5420, phone: 202-762-1412 fax: 202-762-1516 email: bdm@draco. usno.navy.mil.

USNA1.0, 2.0, SA2.0, UCAC1

Results of PMM scans of Schmidt survey plates – USNA2.0 has 550 million stars on 10 CDs. SA2.0 is more manageable – 55 million stars. UCAC1 is the best astrometrically (covers dec –90º to –5º, accuracy = 70 mas at $R = 16$) http://ad.usno.navy. mil/star/star_cats_rec.html

The Webb Society

Regular publication of measures. Advice on observing. There is a comparison between the RETEL and van Slyke micrometers on the website. www.webbsociety.freeserve.co.uk/notes/ doublest01.html.

Christopher Lord

An excellent website containing many useful references for serious double star observers including details of filar screw evaluation, observation of unequal double stars, and a thorough description of the Lyot micrometer and its use www.brayebrookobservatory.org.

Martin Nicholson (www. double-star.org.uk).

Details of recent double star astrometry using a 12-inch Meade LX-200 and SBIG CCD.

Royal Observatory, Brussels (www.astro.oma.be/D2/DSTA RS/index.html)

Especially useful for CCD techniques.

Francisco Rica Romero (Spain)

Some measures (www.terra.es/personal/fco.rica/Dobles.htm)

Observatori Astrónomic del Garraf (www.oagarraf.org)

Spirit of 33 – e-observing group.

Very active group – some good links (www.carbonar.es/s33/33.html)

The *Double Star Observer*

A magazine published in the USA. Contact the Editor, Ronald Tanguay, 306 Reynolds Drive, Saugus, MA 01906-1533, USA.

(e-mail: rctanguay_dso@yahoo.com). Some useful links ttp://home.cshore.com/rfroyce/dso

Societe Astronomique de France (Double Star Section) (www.iap.fr/saf/cometdbl.htm)

The publication *Observations & Travaux* sometimes has issues dedicated to double star observing. The last such issue was no 52.

Asociacion Argentina "Amigos de la Astronomia"

Contact Mauro Gallo (e-mail: maurogallo@hotmail.com) c/o Avenida Practicias Argentinas 550, 1405 Buenos Aires, Argentina

Double Star Section, Royal Astronomical Society of New Zealand

Contact: Warwick Kissling, PO Box 1080, Wellington , New Zealand. Tel: +64 4 5690351 Fax: +64-4-5690003, e-mail: w.kissling@irl.cri.nz.

Hungarian Double Star Section

Contact: Ladányi Tamás, 8200-Veszprem, Fenyves u. 55/A, Hungary, e-mail : lat@sednet.hu.

Double Star Section, Astronomical Society of Southern Africa

Contact: Chris de Villiers, Suite 129, Private Bag X7, Tyger Valley, 7536 South Africa, e-mail: astronomer@skywatch.co.za (www.skywatch.co.za/doublestars/index.htm)

Hipparcos (www.astro.estec.esa.nl/Hipparcos)

Tycho-2 (www.astro.ku.dk/~erik/Tycho-2/)

Astronomical League

Encourages double star observing
(http://astroleague.org/al/obsclubs/dblstar/dblstar1.html)

Alejandro Russo (Argentina)

Traditional techniques of measurement including chrono-metric micrometer (www.geocities.com/CapeCanaveral/Runway/8879/)

Professor W.D. Heintz

Catalogue of orbital elements (laser.swarthmore.edu/html/research/heintzr.html)

Richard Harshaw

Leader of project to assess Micro Guide eyepiece for double star astrometry
(e-mail : dkharshaw@kc.rr.com).

Brief Biographies

Andreas Alzner

After studying physics and astronomy in Bonn, Andreas completed a dissertation in nuclear physics in 1985 and followed this with work in the electrical industry as technical instructor for magnetic resonance imaging systems where he remains.

His early interest in amateur astronomy from 1968 to 1992 consisted of observations with reflectors (4.5-inch, 6-inch, 8-inch, 14-inch) and refractors (5-inch, 6-inch), but (he says) nothing scientific. He was interested in double stars from the beginning on but his telescopes were never good enough for measurement work.

His first really good telescope, a 14-inch Zeiss Newtonian, was acquired in 1992 followed in 1996 with a long-focus 13-inch Cassegrain. Since then he has made several thousand measures with filar and double image micrometers and has also published a number of orbits in *Astronomy and Astrophysics* and the Circulars of IAU Commission 26.

Graham Appleby

Graham Appleby spent his working life on various projects at the Royal Greenwich Observatory in Herstmonceux and at Cambridge until its closure in 1998. At that time he transferred to the Natural Environment Research Council where he continues to work within the Space Geodesy Facility. He has a Mathematics BSc and an Aston University PhD in Satellite Laser Ranging. Graham has long been interested in the lunar occultation technique, having made a large number of visual observations and carried out various scientific analyses. He is currently involved in using the SLR system to make high-speed photoelectric observations of occultations for double star and stellar diameter determination.

Bob Argyle

His blinkered interest in double stars dates back to the late 1960s and a period at the Royal Greenwich Observatory (RGO) at Herstmonceux in 1970 when he was let loose on the 28-inch refractor only made it worse. Occasional and all-too-short periods of observing occurred until 1990 when the RGO moved to Cambridge and Bob along with it. The availability of the 8-inch refractor satisfied a long-desired need for regular observation which is still in progress today. Bob works at the Institute of Astronomy where he is responsible, amongst other things, for the distribution of archive data from the Isaac Newton Group of Telescopes on La Palma, and where he spent a period as a support astronomer from 1984–1988. He is a member of Commission 26 (Double Stars) of the International Astronomical Union and an Editor of *Observatory* magazine. He is President of the Webb Society and has directed the Double Star Section since 1970.

Owen Brazell

As well as editing the Webb Society *Deep-Sky Observer*, Owen is also the assistant director of the British Astronomical Association's Deep-Sky Section and a regular contributor to *Astronomy Now*. When observing, his primary interests are in the observation of planetary and diffuse nebulae – although since the acquisition of a 20-inch Obsession telescope this has also moved to viewing galaxy clusters. His interest in astronomy was sparked by an attempt to see a comet from his native Toronto. From early years, he kept up his interest in astronomy which culminated in a degree in astronomy from St Andrews University in Scotland and taking though not completing an MSc in Astrophysics. At that time, he also gained an interest in the northern lights. As with many astronomers, finding no living there, he moved into the oil business first in R&D and then as a computer systems designer (this explains his interest in the computer side of astronomy). Despite this he still uses Dobsonian-type telescopes ranging from a 4-inch Genesis-sdf up to the Obsession. The recent plethora of fuzzy objects that move has re-awakened an interest in comets! His searches for dark skies have taken him from the mountains of Canada through Texas to the Florida Keys as well as to Wales – the only good dark sky site he has found so far in the UK.

Michael Greaney

Michael Greaney is a member of the Webb Society and an award member of the Auckland Astronomical Society. His astronomical interests lie mainly in the field of spherical astronomy, with a particular interest in double stars. He has written a number of articles on the subject which have appeared in publications in the UK, the USA and New Zealand. Outside of astronomy, Michael works as a domestic manager (looking after house and home for his working wife and three school children).

Andreas Maurer

Andreas Maurer is a mechanical engineer and a lifelong astronomy enthusiast. Since his recent retirement, he is now able to concentrate on his astronomical interests. Besides activities related to the history of astronomy he is building his own telescopes and is restlessly experimenting with home-made auxiliary equipment suitable for amateur observations. Whenever nightly seeing conditions are favourable he observes double stars from his home in Switzerland.

Michael Ropelewski

Mike Ropelewski is an active member of the British Astronomical Association and the Webb Society. His main interests are the study of aurorae, comets, double stars and eclipses. His instrumentation includes 15 × 45 stabilised binoculars, a 102 mm SCT and a 250 mm Newtonian reflector in its own observatory. In 1999, the Webb Society published his first book entitled *A Visual Atlas of Double Stars*. During daylight hours Mike is a computer programmer/analyst by profession. Apart from astronomy, he enjoys gardening, music, poetry and steam railways.

Christopher Taylor

Originally trained as a theoretical physicist, Christopher Taylor teaches mathematics and astronomy over a wide range of undergraduate courses and is tutor on the University Department for Continuing Education's long-running astronomy evening classes in Oxford. He is

Director of the Hanwell Community Observatory, a public educational venture set up in partnership with the Oxford Department under the Royal Society's Millennium Awards Scheme. This will contain one of the largest telescopes in Britain wholly dedicated to public and educational astronomy, as well as other instruments from 4 to 30 inches aperture (0.1 to 0.76 m) available for amateur research. Christopher Taylor has been an active observer since 1966, for most of that time using the same 12.5-inch (0.32 m) reflector, with a long standing interest in visual binaries which has become his main observational pursuit since 1992. Other observational interests are high-resolution optical work in general (including, e.g. planetary), optical spectroscopy and broadly anything quantitatively measurable in the sky. For further information on the Hanwell Observatory see (http://www.hanwellobservatory.org.uk)

Tom Teague

Tom Teague is a Fellow of the Royal Astronomical Society and a member of the Webb Society and the British Astronomical Association. He has written articles for *Sky and Telescope*, the *Journal of the British Astronomical Association* and the *Webb Society Quarterly Journal*, covering such topics as double-star micrometry, sunspot measurement and amateur spectroscopy.

Nils Turner

Nils Turner has been using speckle interferometry to observe binary stars on large telescopes since 1990. Since 1996, he has used adaptive optics to study binary stars, concentrating on relative photometry as opposed to astrometry. He is a member of the American Astronomical Society. By day (and night), he works in the field of optical/IR Michelson interferometry. Away from astronomy, Nils enjoys Linux programming, playing viola in a community orchestra, playing Ultimate (frisbee), and spending time with his wife (not necessarily in that order).

Doug West

Doug West is an active observer of double stars, variable stars, and Solar System bodies. He is a member of the American

Astronomical Association, American Association of Variable Star Observers, and the International Association of Photoelectric Photometrists. His observations and analysis have been published in the Webb Society *Double Star Circulars, Double Star Observer* newsletter, *Bulletin of the American Astronomical Association*, Association of Lunar and Planetary *Observers Bulletin*, and in the MidAmerican Astrophysical Conference. Doug is an aerospace engineer during the day. Outside of astronomy, his interests include gardening, being a soccer coach, and he is a worker in his church's youth group.

Index